Science and Inequality

We dedicate this book to those who are next in line.

Scott: to Eva and her cousins, Devin, Rebekah, Justin, Hunter, Lucy and Finley

Kelly: Amari, Antonia, Evelyn, Gavin, Jamie, Riley, Río. Silas, and Zoran

Science and Inequality

A Political Sociology

Scott Frickel
and
Kelly Moore

polity

First published in 2025 by Polity Press

Polity Press
65 Bridge Street
Cambridge CB2 1UR, UK

Polity Press
111 River Street
Hoboken, NJ 07030, USA

ISBN-13: 978-1-5095-1479-3 (hardback)
ISBN-13: 978-1-5095-1480-9 (paperback)

A catalogue record for this book is available from the British Library.

Library of Congress Control Number: 2023934599

Typeset in 10.5/12pt Sabon LT Pro
by Cheshire Typesetting Ltd, Cuddington, Cheshire
Printed and bound in Great Britain by Ashford Colour Ltd

For further information on Polity, visit our website:
politybooks.com

Contents

Acknowledgments

Science and Inequality is a very small book about an unfathomably large problem. It began ages ago with a "what if" conversation with our editor, Jonathan Skerrett, in 2011. Today, we owe him our deepest thanks for the enduring patience he has shown us over the intervening years. In fits and starts, as different versions of our argument came together, fell apart, then came together again (often with long lapses in between), Jonathan provided forthright guidance and consistent encouragement. He helped us maintain our faith in a project that was, at best, depressing and at times – through the (first) Trump years, Covid-19 pandemic, the U.S. Supreme Court's toppling of affirmative action and reproductive rights, and all the rest – difficult to see our way through the "wine-dark seas"[1] of contemporary U.S. and world politics.

Scott: Whatever insecurities we carried for the project were assuaged along the way by family, friends, and colleagues who read and commented on the drafts, partial drafts, and chunks of text we sent them. For their critical feedback, insights, suggestions, and conversations, I thank Ale Hannud Abdo, Florencia Arancibia, Grace Berg, Beth Fussell, Meredith Hastings, Meghan Kallman, Adi Ophir, Tara Nummedal, Jonathan Skerrett, Laura Stark, Nathalia Hernández Vidal, Rhys Williams, and two very astute and generous anonymous reviewers. I thank Katy Pickens and Brianna Nee, who provided excellent support as undergraduate research assistants, and Susan Beer for sharp copy editing. To the broader STS community of scholars, I am grateful for your individual and collective insights and hard work, which has nurtured and sustained me in this project and over the years. David Hess and Daniel Kleinman, in particular, have long offered safe harbors to try out new ideas and tinker with old ones, usually over a wonderful meal. In the end, we alone are responsible for this little boat, whether it sinks or swims.

Kelly: My thinking about science and inequality and the possibilities for science to deliver on its promise for better life for the bio-social world has been shaped by many people. I thank Savina Balasubramanian, Elizabeth Bernstein, Nathalia Hernández Vidal, Miliann Kang, L.A. Kaufmann Kerwin Kaye, Daniel L. Kleinman, Anna Linders, Jackie Orr, Shobita Parthasarathy, Madeline Pape, Gwendolyn Purifoye, Joseph Renow, Sigrid Schmalzer, Laurel Smith-Doerr, Bruno Strasser, Banu Subramaniam, Alice Weinreb, and Kyle Woolley for comments and discussions that deepened and enriched my thinking on these topics. Audiences at Northwestern University, Michigan State University, the Annual Meeting of the American Sociological Association, the Annual Meeting of the Society for Social Studies of Science, and the Purdue University Conference on the History of the Cold War provided critical insights and opportunities for trying out new material. The University of Massachusetts-Amherst Department of Sociology and Department of Women, Gender, and Sexuality Studies provided intellectual and material support and new ways of thinking during a 2019 sabbatical. For his steadfast support for this project, and his rich and insightful guidance throughout its development, I am deeply grateful to Jonathan Skerrett. Many thanks to the staff at Polity Press for their expert work on the production of the book.

To our pole stars, Beth and Rhys, as always, thank you for everything.

1

Science, Society, and the Paradox of Inequality

Long associated with human progress, economic prosperity, and social betterment, for most people today science is a quintessential public good, the *sine qua non* example of human ingenuity and cultural achievement. The more knowledge that scientists and engineers make available to the world, the better off are the planet and its human inhabitants. Two fundamental and related ideas embed this common belief.

One is the idea that science and its products are uniquely powerful in providing a foundation of knowledge about the world and of humans' place and potentialities within it. This idea is so commonly and deeply held that for many people it is all but invisible. People learn to trust in the social value of science from an early age and our daily dependence on the products of modern science constantly reinforces this trust – even among skeptics who criticize what science tells us about certain things, like climate change, dinosaurs, vaccine efficacy, or sexuality. Belief in the social value of science is expressed consistently in national and international public opinion surveys (National Science Board 2022a; Haerpfer et al. 2022), and helps explain why general education requirements for college baccalaureate degrees include courses in math, chemistry, biology, and physics, and why Nobel Prize winners garner annual worldwide media attention. More prosaically, we believe in science because science works. Every day, people board airplanes, take medicine, and log onto their computers, routinely trusting in the power of science to help us travel, heal, connect, learn, and live.

The second idea embedding popular belief in the social value of science is that the more knowledge scientists and engineers make available to the world, the less divided and unequal the world becomes. *Prima facie* evidence for this idea is perhaps best reflected in public education systems that provide children around the world with free access to basic scientific and civic knowledge. Other kinds of sociotechnical systems illustrate a similar vision: cars and highway systems, satellites and social media, the International Space Station, and the Human Genome Project.[1] These are just a few examples of the ways in which science and engineering ostensibly build shared understanding around a common bank of knowledge and technologies that bring people closer together and lift up a progressive vision of society based on opportunity, equality, and justice. Embraced by economists and policymakers in particular (e.g. International Monetary Fund 2021; Kim and Quresh 2020), the Enlightenment-era idea that science, liberty, and equality rise together is institutionalized in legislation that establishes national funding priorities for science and sets scientific research agendas, and in grant proposals that scientists write to secure research funding. It is embedded in legal statutes granting corporations rights to patent new life forms and protect trade secrets and proprietary information. It is a feature of global economic indicators such as those found in United Nations *Human Development Reports* (UNHDP 2022) and many popular accounts of scientific heroism, discovery, and invention. It is an overt aspirational goal of many academic fields, from agroecology and development economics to social work and tropical medicine, and it is the impetus behind academic and popular enthusiasm for participatory science projects now sprouting by the thousands in communities around the world (Irwin 2018).

Science and Inequality challenges both of these ideas. The popular belief in the unique power of science to increase social equality often ignores how power actually operates within the field. Our book's central assertion is that inequality is built-in to science because, while not every scientific field and research topic is contoured by power in the same ways, all science is an enduring product of power and its unequal circulation, use, and impact. The consequences of technoscientific power are uneven, too. Built-in inequality in science has profound implications for human societies and the world. Yes, science works, but for whom? Where and under what conditions does it work? And to what ends? Of course, science can and does address important problems in the world, and deserves our trust. But our trust in science should remain cautious, not blind.

For example, science can attend to important problems in ways that also render societies less equal, or that consistently benefit certain groups more than others. Studies of variability in patterns of gender-based inequality in the global scientific workforce (Zippel 2017) show that in some countries, women make up more than half of that workforce, and in other countries they make up a tiny percentage. As Anthony Hatch (2022) has shown, in the context of the Covid-19 pandemic's impacts on communities of color, medical researchers' repeated study of certain topics in the absence of corresponding action to address the problem of disproportionate morbidities can operate as scientific gaslighting. He writes (2022: 2): "the painstaking calculation of the pandemic's disproportionate effects on racial and ethnic minority groups is a necessary but insufficient condition for saving those lives." Scientists' repeated efforts (over decades prior to the onset of the pandemic) to statistically document racial health disparities paradoxically legitimates and reinforces a medical research and health care system from which racial and ethnic groups receive disproportionately less attention and lower quality care.

In other ways, scientific research can ignore marginalized groups, exemplified by the lack of research on environmental health problems that disproportionately affect the poor, Indigenous, and other minoritized groups around the world (e.g. Fernández-Llamazares et al. 2020; Pickett and Wilkinson 2015; Prüss-Ustün et al. 2016). Research can also exploit marginalized groups to gain access to commercially promising resources, illustrated in cases of bioprospecting for Indigenous genetic material (Hayden 2004). In still other ways, research can profit from poor communities lacking the political tools and resources to resist harm from industrial or other hazards (Auyero and Swistun 2009; Davies 2018).

We find epistemic inequality in the scientific racism and sexism promoted by biologists, anthropologists, and psychologists centuries ago (Gould 1981; Hammonds and Herzig 2009; Schiebinger 1993) and in today's new genetic and genomic approaches to human population variability (Bliss 2012; Duster

1991; Reardon 2004; TallBear 2014). We also find it in the social histories of concepts, categories, and measures that social scientists and neuroscientists use to study poverty and inequality today (Hirschman 2021; Pitts-Taylor 2019; Rodríguez-Muñiz 2015), and in the uneven distribution of the benefits of science and technology (Bridges, 2011; Benjamin 2013; Noble 2018) that simultaneously cause and perpetuate inequalities. In these examples and more, patterned inequalities coursing through science undermine the long-standing expectation that harnessing reason through methodologically rigorous study will lead inexorably to increased political freedom and spread economic benefits to more individuals and societies (Merelman 2000).

This book advances a framework and methodological guidelines for identifying patterns of inequality that emerge from the consideration of many different contemporary and historical cases, like those noted above. By investigating science's relationship to inequality and why science – an institution that explicitly incentivizes and rewards novelty and innovation – so often proves stubbornly resistant to change in its power structures, we hope to gain deeper understanding of how science operates in the modern world. In this way, a political sociology of science might nudge scientific institutions toward a more equitable distribution of technoscientific power, reducing certain kinds of inequality and allowing life, for specific groups and more broadly, to flourish.

Why We Wrote this Book

The impetus for writing *Science and Inequality* emerges from two related observations. The first is that ubiquitous reliance on science and engineering to sustain and advance the ecological, material, and cultural bases of human life and well-being parallel deep and growing divisions within and between societies. As the brief examples in our opening section allude, neither the clear benefits of science nor its real and potential harms are distributed evenly, meaning that social inequalities are intensifying even as social investments in expert knowledge multiply.

Similar divisions and inequalities also exist *within* science. For example, science journalists now talk about "science's 1%," shorthand for alarming increases in salary-based inequity among academic researchers (Lok 2016). The same term references the fact that about 150,000 scientists around the world author 87% of global scientific publications (Ioannidis, Boyack, and Klavans 2014), suggesting that an intertwined and steeply hierarchical system of knowledge production supported at the top by grants and systems of promotion and profit are baked in to the structure of elite science. At the bottom are a depressing litany of statistics from higher education where, for example in the United States, Native American college and post-graduate enrollment and degree attainment – already disproportionately underrepresented among college and graduate school students – have been declining for the past decade (Field 2016; Postsecondary National Policy Institute 2021), while Native American representation among STEM faculty "in all [sampled] disciplines ranges from miniscule to zero" (Nelson and Madsen 2018: 382). For academic researchers in our own field of science and technology studies (STS), growing inequality amid unrivaled scientific and technological achievement emerges as a central paradox of our time. Inequalities proliferate across the worlds we study and the worlds that we inhabit as STS researchers (see Harding 2002; Mascarenhas 2018).

A second reason for writing this book is that while examples of social and environmental inequality are all around us, available in news stories or culled from our own biographies, these problems can often seem disconnected and unrelated to science and to one another. Thus, we may encounter a front-page newspaper article about how pesticides simultaneously increase crop yields but also sicken farmworkers, and a few pages later read a different story about advances in fertility treatments that disproportionately benefit expectant parents with higher incomes compared to expectant parents who are poor. Most readers will likely conclude that because the two news stories address different substantive problems – fertility and pesticides – the causes and forms of inequality in each must be singular and distinct. We do not often think to seek connections between food systems that are heavily reliant on synthesized chemicals designed specifically to take invertebrate life, and decreasing fertility rates among vertebrates, including humans (Colborn, Dumanoski, and Meyers 1997). But they are connected through science.

STS scholars often fail to make the connection, too. For decades, mainstream currents in STS paid relatively little attention to social inequality, especially relative to bright-light topics involving new discoveries, technologies, and expertise at the forefront of fast-growing research fields, such as communications and surveillance, nanotechnology, data science, and genetics (among others and which are, themselves, implicated in the production and reproduction of inequality). As a field, STS has historically failed to study science's relationship to inequality *per se*[2] and thus we have limited understanding of how presumably unconnected problems of science – like rising rates of infertility and agricultural use of endocrine-disrupting chemicals – might reflect common patterns, linkages, and origins in the sociopolitical world.

Yet, in the past decade, a new wave of STS research led by feminist, Indigenous, critical race, and Global South scholars has moved the question of science and inequality toward the center of STS. The cross-field and multi-institutional reach of many of these intersectional studies offer novel analyses of race-, class-, gender-, sexuality-, and ableist-based inequalities that derive from or in other ways implicate science.[3] These studies spotlight the consequences of technoscientific power on marginalized or otherwise underrepresented groups and develop explanatory frameworks that draw connections across social and cultural structures such as race, gender, and Indigeneity (Hatch 2016; Liboiron 2021; Murphy 2017a), that can help us find commonalities and trace patterns across apparently separate and unconnected processes and events. These scholars also design their scholarship with an eye toward real-world impact by working with communities to address urgent concerns of resistance, refusal, repair, and regeneration (Benjamin 2019). Their collective insights inspire and inform this book.

Our own earlier effort to advance a political sociology of science raised similar concerns about inequality and power in the formulation of scientific problems, research methods, and the distribution of advantage and disadvantage (Frickel and Moore 2006a).[4] Formulated in response to what we then viewed as overly empiricist, consciously apolitical and deconstructive approaches to science that had long dominated STS (many of the core arguments are cataloged in Pickering 1992), we argued that a distinctly sociological set of theoretical and methodological tools were needed to unpack science in relation to sociology's core concerns with power and inequality. Where constructivist STS looked for and inevitably found theoretically interesting cases, we were interested in identifying similarities and differences running *across cases* that would reveal larger patterns and lead us to more general conceptual, theoretical, and practical tools, and claims about how science works and for whom. Our goal then was to encourage research into how unequal power relations convert into scientific knowledge and authority. Two decades later we find ourselves in more plentiful scholarly company, even if our goal and strategy remain much the same: to call inequalities within science and those created by science out from the shadows by plumbing a range of disparate cases for common causes and general insights, the better to extend shared knowledge and collective practice on processes of repair, remediation, and harm prevention.

The examples we examine throughout *Science and Inequality* differ in many ways and we have no interest in flattening out those differences to fit them artificially into an overly simplistic causal framework. Rather, our intention is to analyze them as a class of related phenomena, or as elements or features of a social formation – more specifically, a *scientific inequality formation*. This idea is our organizing framework. Later in this chapter, we describe and illustrate this framework, which we use throughout the book to gain analytic purchase on inequality and its complex relationship with science and scientific knowledge. To set the stage properly, we first need to lay out some basic terms.

Inequality, Science, Values, and Politics

Our analysis of scientific inequality formation features four central concepts. The first is inequality.

By **inequality**, we mean the uneven allocations of societal benefits and risks that can generate starkly different capacities and experiences among individuals and social groups to thrive and steer the course of their lives. Particularly important to us are those "durable" inequalities that are sustained over longer periods and have a stickiness that is difficult to dislodge because they are embedded in the routines, discourses, and practices of daily life, in legal and administrative systems and across interlocking institutions (Tilly 1998). Here, we are thinking of societal inequalities as they are structured primarily by wealth, health, education, citizenship status, race, ethnicity, sexuality, gender, and ability. We can often find complements to these sorts of durable inequalities within science. For example, the U.S. gender wage gap in science mirrors salary inequalities between female and male workers generally (Smith-Doerr et al. 2019), while "Blacks, Hispanics, and American Indians or Alaska Natives collectively represented 30% of the employed U.S. population but 23% of the total STEM workforce in 2019 (National Science Board 2022b).

We find other durable inequalities embedded in the categories that organize scientific disciplines. For example, despite many scientists' insistence that race is not inherent in human biology, older and explicitly racist ideas about human difference and racial hierarchy remain stubbornly entangled with newer scientific discourses in population genetics. Even newer genomic sciences which claim to move "beyond race" in offering "color blind" analysis of human population differences continue to be influenced by older ideas of racial difference (Bliss 2012; Fujimura and Rajagopalan 2011; Nelson 2016). As Aaron Panofsky and Catherine Bliss argue (2017: 61), the tendency to label human populations based on continental geography[5] rather than using traditional racial categories to distinguish populations (e.g. "African" rather than "Black"), "ambiguously blends racial and geographic ways of conceiving populations." These newer geographical conventions produce a type of epistemic ambiguity that prioritizes geneticists' views over other actors – be they competitors, study participants, funders, or social justice movements – all of whom may implicitly theorize human genetic distinctions in very different ways. In this example, we see how shifting social values within science are evinced in new terminology that nevertheless reinforces the durable hierarchies of racial difference as legitimated within presumptively post-racial genetic sciences. The example also affords a view of science's lack of reflection in its historical role in propping up race- and ethnicity-based social disparities.

By **science** (and occasionally the intentionally blurred "technoscience"), we mean the organization and shared processes of systematic knowledge production, or research, that aims to understand the functioning of natural and social systems. The research process and the organizational infrastructure that facilitates the production and circulation of scientific knowledge (such as universities, corporations, private foundations, and regulatory agencies) as critical sites for the (re)production of, but also epistemic challenges to, durable social inequalities. Our definition includes social sciences as well as physical, chemical, life, and engineering sciences. An important arena where such inequalities play out involves disputes over knowledge that challenges powerful interests, a common dynamic in many environmental and health conflicts.

For example, in 2010, Argentinian medical students and faculty at a regional university began working with allies from grassroots movements to routinize new epidemiological data collection practices designed to call medical students' attention to the public health consequences of industrial agriculture's indiscriminate use of pesticides. Despite using these new tools to collect and distribute much-needed epidemiological data for more than 100,000 residents in forty impacted communities and growing recognition from national and international media, the project sparked increasing resistance from large agricultural firms that used pesticides. As a result, in 2019 university authorities summarily, and without warning, shuttered the project (Arancibia et al. 2022). As this case suggests, disputes over research involving collaborations with scientists and communities often have less to do with the social utility of knowledge, than with the historical and institutional power differentials between those who most need certain kinds of knowledge and those who have traditionally controlled its production (Kimura and Kinchy 2019).

Values are shared normative understandings of the world about things we hold to be socially important, desirable, proper, or worthy. Values can involve abstract ideals, morals, or principles but can also take material expression in physical objects, infrastructure, organizations, or social roles and practices. Social values are important because they shape peoples' interests and beliefs and guide individual and social action. For example, indications that a society values education may be that governments create tax policies to fund public schools, families put away savings to help pay for children's college, and students experience peer pressure to perform well on homework and exams. Famously, Max Weber (1949) held values to be inherently nonrational, in contrast to scientifically derived facts, in calling for an empirically descriptive sociology that strives for value-freedom, as he believed all sciences should. Weber's ideas about the incompatibility of facts and values (which is itself an expression of social values) are common today and is one reason why values in science so often go unexamined.

STS generally rejects Weber's formulation, insisting that facts and values cannot so easily or straightforwardly be separated and that historical debates over science's role in society have involved precisely the kind of value-oriented arguments and considerations that Weber (and, after him, Robert Merton) saw as antithetical to science (Gieryn 1999). We find a way forward in Abby Kinchy and Daniel Kleinman's (2005) contention that because science is not and never has been free of values, it is imperative that social scientists investigate the value-orientations that emerge from science. Doing so informs public debate on difficult conversations about *which* values, among many, best align scientific practices with societal needs for new knowledge. As we will argue throughout this book, it is precisely the absence of routine attention to the value-composition of science that has generated conditions for scientific inequality formation as an enduring feature of modern society.

By **politics** we mean, following political scientist David Easton (1965), the "authoritative allocation of values." Politics consists of the processes by which groups use their authority to organize society and distribute public goods and resources in ways that reflect and reinforce certain value sets – including those embedded within science. Power (conferred as social authority) and morality (as described by social values) are thus inherent in politics, no less so in the politics of science. We part with Easton and many other political scientists in formulating politics primarily as a feature of formal government.[6] We certainly find the politics of science in legislative chambers and executive offices and in administrative extensions of state power such as courtrooms and regulatory agencies. But the politics of science is also present at protest marches and in community science projects as well as in university classrooms, academic conferences, industrial laboratories, or far-flung field sites. And we find it in the endemic relations of racism, classism, colonial legacies, and other structural forms of power that are woven through scientific questions, techniques, and systems of circulation, recognition, and credibility.

Applied to science and inequality, this definition of politics allows us to examine how technoscientific power operates within scientific routines and rules that are legitimated in scientific practice wherever it occurs, including by structuring the production and non-production of knowledge, as we show in Chapter 3. This definition of politics also helps us focus attention on profit-making (the subject of Chapter 2) as a dominant value in science that reinforces social and scientific hierarchies. This definition also provides a way to investigate the potential for structural and cultural transformations in science, rooted in changing distributions of values. Whether and how values are being altered and realigned to distribute technoscientific power and resources more equitably is the subject of Chapter 4. The socio-historical result of these ongoing, often intertwined, processes, is scientific inequality formation.

Scientific Inequality Formation

Scientific inequality formation is our take on an old sociological idea. In *Capital*, Karl Marx wrote about "social formations" to describe the slow, macro-historical shifts in economic and political organization that helped him explain how ancient societies transformed into feudal societies and how feudal societies then transformed into capitalist societies. Marxist scholars in the twentieth century further developed the idea of social formation, showing how cultural representation or ideology reproduces the conditions of economic production and political domination. In feudal societies, for example, the church played central ideological functions in maintaining the status quo; in capitalist societies, this function is distributed more broadly among other systems, including the media and education (see Hindess and Hirst 1977).

Sociologists Michael Omi and Howard Winant further developed the idea of social formation with their theory of "racial formation." They defined racial formation as "a socio-historical *process* by which racial categories are created, inhabited, transformed, and destroyed" (1994: 55; our emphasis). For them, racial formations are built from "racial projects" or collective efforts to simultaneously construct racial categories and "reorganize and redistribute resources among particular racial lines" (56). In this way, racial projects "connect what race *means* in a particular discursive practice and the ways in which both social structures and everyday experiences are racially *organized*, based upon that meaning" (ibid.; emphasis in original).

Seen this way, racial formations are expressions of dominant and subordinate social values, materialized in rules and routines and challenges to those rules and routines. Jim Crow laws, the freedom movements of the 1960s and the Movement for Black Lives founded in 2015 are examples of racial projects that, in combination with other racial projects, have reinforced or altered racial formations in the United States since the end of the Civil War. In Colombia and in other parts of Latin America, racial projects since the time of Spanish colonization took the form of considering all citizens mestizos, or "mixed," obscuring the ways that scientific claims and structures of political and economic power left Black people in worse conditions than those of lighter-skinned people. Like other racial projects, this one is also changing: the 1991 Colombian constitution gives rights and recognition to Afro-Colombians, setting aside "ancestral lands," while simultaneously asserting that other territory is open for mining and agribusiness (Asher 2009).

A few years later and building on Omi and Winant's work, environmental sociologist David Pellow (2000) advanced a theory of "environmental inequality formation." Pellow's theory draws attention to the important idea that *"inequalities are relationships* that are constituted through a continuous process of change that involves negotiation and often conflict among multiple stakeholders" (Pellow 2000: 589; our emphasis). This means that inequalities are not given *a priori*, but emerge over time through ongoing social interactions. Social interactions may occur among social groups, as between scientists, policymakers, and laypersons, or between institutional orders, logics, and discourses as we see in cases where regulatory science is challenged in court by citizen's groups or industry trade associations (Jasanoff 1990). For example, in one of Pellow's earlier studies, about the so-called "garbage wars" in Chicago (2002), environmental inequality is contested but ultimately reinforced through a series of conflicts and policy initiatives involving city managers, commercial waste management companies, neighborhood residents, and employees at a local recycling and waste sorting center.

Pellow's focus on inequality as a historical process and his insistence that we take seriously the relational nature of inequality is broadly consonant with our own. We take this to mean, for example, that we should not study cases of environmental conflicts between residents of low-income communities and industrial polluters as discrete events or outcomes, but as moments unfolding across larger, ongoing socio-historical processes that, across diverse and seemingly disparate local cases, generally favor the interests of wealthy industrial polluters over the interests of working-class communities (Taylor 2014).

For this reason, Pellow encourages his readers to look for connections among cases – or projects, using Omi and Winant's term – to understand how they fit together and build to something more durable, if often less visible. For Pellow, this means taking a lifecycle analysis approach to environmental injustice research, following pollutants from their points of extraction as raw materials, to their transformation in factory production, to consumption and waste disposal. By following the material objects in this way, students of environmental inequality can begin to piece together how people living in far-flung places and who experience pollution in different ways are connected by an environmental inequality formation that prioritizes profit and growth over human and ecological well-being.

We can turn many of Omi and Winant's and Pellow's conceptual insights and methodological sensibilities toward the formation we are studying. The first thing to note is that all sciences are in the business of creating categories that organize human understanding. Thus, science is organized by ongoing processes of differentiation and integration among different projects to define what counts as scientific knowledge (rules) and how to count it (routines). Weights and measures, the periodic table, and biological taxonomic trees are all common examples, as are many categories through which we understand social inequality: race, gender, class, ability, sexual orientation, national or post-colonial identity, and the like. The development and use of inequality categories (e.g. annual household income) and social groups (e.g. "the poor") are cultural projects of scientific representation and thus political insofar as scientific concepts codified in rules and routines are expressions of the allocation of values and resources in science and society. In the framework we are developing, cultural projects to define categories of knowledge become sites of epistemic conflict as well as the stakes of struggle over what counts as science, who gets to make science, and how the resulting knowledge circulates in broader society. We view these dynamics – which may at times be contentious and at other times quiescent – as episodes unfolding across larger, ongoing processes of scientific inequality formation.

Science is also in the business of creating socio-material and digital systems, or infrastructures (Slota and Bowker 2016; Hamrai 2017). From oil refineries and pipelines to college testing and admissions processes, infrastructures are configurations of technology, knowledge, and expertise that bake-in social ties and practices, stabilizing inequalities in ways that make them difficult to dislodge or undo. For example, decisions about new energy pipeline construction may appear at first glance to be based on purely technical criteria, such as topography, hydrology, soil composition, and characteristics of a particular fuel type. Yet, debates about gas and oil pipelines in Canada, Nigeria, the United States, and many other countries show there is nothing pure or apolitical about pipeline science and engineering (e.g. Whyte 2019). Technical answers to questions about whether and where to build them, and with what safety and cultural protection criteria, are deeply value-laden. Moreover, once the pipelines are built, the new energy infrastructure materially cements a society's commitment to fossil fuel use for decades. The enormous costs and complexity of pipelines and their inputs and outcomes give scientists, corporations, and governments incentives to keep using them, rather than seek other, non-fossil energy sources. Sold to the public as a safe, effective, and efficient way to secure access to oil and natural gas – and, more recently as a necessary "bridge" to non-fossil energy sources (Delborne et al. 2020) – pipelines and related energy infrastructure offer a literal illustration of politics as the "authoritative distribution of values" (Easton 1965) etched across landscapes, economies, national borders, and cultural spaces.

A different example involves the monitoring infrastructures that labor economists and social statisticians developed in the 1930s and 1940s to study income inequality – then conceptualized as a problem of poverty. As Daniel Hirschman (2021) argues, this framework in part blinded them to long-term increases in top income earners beginning in the 1980s and who became infamous in the 2000s as "the 1%." Such outcomes are shaped in part by science's significant contributions to economic growth, enacted through a vast research and development apparatus that contributes to the economic and social inequalities and instabilities that we have seen, for example, in advancing the cause of war-making in the United States and around the world (Moore 2008; Rohde 2013, 2017). Like pipelines, datasets, statistical analyses, and the policies built from them also embed values in ways that science structures social advantage and disadvantage.

Our larger point is that at any given moment in time, various cultural and structural projects are either reinforcing, dismantling, or otherwise altering knowledge production systems. These projects unfold in distinct historical and social contexts, involving multiple groups and actors whose work collectively defines the world and sets out various courses for scientific action or inaction. Whether they are contested or quiescent, visible, or submerged, these projects are the building blocks of scientific inequality formation, an ongoing political process of value allocation measured in facts, artifacts, publications, careers, and routinized practices that condition what we know about the world and what remains unknown, undone, unheeded, or unseen (Hess 2016; Hoffman 2013).

An example of the latter is provided by Lauren Richter and colleagues (2018), who examine the misalignment between decades of extensive industrial research on the toxicological effects of per-and polyfluorinated alkyl substances (PFAS), and the relatively recent recognition of PFAS as a dangerous class of "emerging contaminants" by government regulatory agencies (see also Wickham and Shriver 2021). Why, they ask, has scientific knowledge about the ecological and health impacts of these ubiquitous chemicals provoked such limited and long-delayed regulatory action? As their study helps to illustrate, scientific inequality formation can operate by compartmentalizing or sequestering certain kinds of knowledge to limit its circulation and thus delay societal responses. As such, communities impacted by PFAS contamination not only encounter scientific inequality formation directly and visibly in courtrooms, or town-hall meetings or at physical sites where contamination exists, but also in less visible systems of obdurate, taken-for-granted ideas, routines, rules, and relationships that allocate technoscientific power through what sociologist Erin Cech and colleagues (2017) describe as "epistemological dominance" – a structural tendency to reject and silence subordinate knowledge, techniques for knowing, and knowers themselves.[7] Consequently, we see scientific inequality formations as almost uniquely conservative among modern social formations; changing them in meaningful and lasting ways requires persistent, coordinated collective action across cultural and structural projects. An example, drawn from the research of sociologist Nathalia Hernández Vidal (2022; see also Hernández Vidal and Moore 2022), will help to illustrate what we mean.

Columbian GM Agriculture as a Scientific Inequality Formation

In 2004, a coalition of Colombian farmers, social and agricultural scientists, and attorneys created the Res de Semillas Libres de Colombia (the Network of Free Seeds of Colombia, or the RSL) to contest a new federal law outlawing commercial use of all traditional or non-genetically modified (GM) seed varietals. The legislation, known to Colombians as Law 9.70, was the latest project in a regional scientific inequality formation – dubbed "neo-extractivismo" by its critics (e.g. Svampa 2015) – that has enshrined industrial scale agriculture, aquaculture, mining, and logging as a dominant Latin American development strategy. The law in question reflected the interests and coordinated efforts between Colombian government officials, Agricultural Ministry scientists, the World Bank, GM seed multinationals, and leaders of other governments, all of whom were committed to the industrialization of Columbian agriculture. These actors designed, produced, and distributed GM seeds, which they viewed as laboratory-based "technical" innovations that promised something for nearly everyone involved. Geo-engineered seeds would improve investor profits, lower loan rates, boost national economic growth, lower consumer prices, and further integrate Colombia into the global finance, biotech, and agricultural product industries that are expanding around the globe.

Left out of the legal equation were legions of Colombia's small-holder farmers, who were caught in a bind. Few had the funds to buy privatized seeds every planting season. On the other hand, continuing to plant traditional (non-GM) varietals meant breaking the law and risking having their crops burned by military decree by agriculture inspectors, incurring large fines and jail time. Faced with the dilemma, many farmers who could not afford GM seeds sold their land, often to brokers eager to consolidate small plots into larger ones for plantation-style agriculture leased to multinational agribusinesses. In the process, farmers became poorer while agribusiness corporations, large landholders, and government officials gained substantial wealth and power.

The consequences of farmer land loss were cultural and epistemological as well as financial. GM seeds and the legal arrangements requiring their use has threatened the stability of local farming practices and subsequent erosion of traditional ecological knowledge. Rather than relying on a bank of community-based knowledge – including knowledge derived from scientific studies – that is shared inter-generationally and reinforced through seed fairs and agricultural celebrations, multinational agribusiness corporations have produced epistemic absences by sending specialists to rural communities to teach farmers how to monitor soil quality, manage irrigation regimes, apply fertilizers and pesticides, and time their harvests for market efficiency and to grow shareholder profits. Of course, some farmers have accommodated these changes. But for many Colombian farming communities, who view seeds as the organizing principle of their traditional lifeways rather than as proprietary food-production technologies, GM agriculture increasingly represents an existential threat. It is a scientific inequality formation insofar as the ongoing struggle is deepening economic inequalities in the Colombian countryside and generating new forms of epistemic inequality between dominant and subordinate knowers.

With the stakes so high, many reject GM seed-based strategies of food production, and instead promote agroecological approaches that organize agriculture in ways that generate cultural, economic, and land vitality, and give power to people who grow food (Altieri 1995). In this context, RSL is fighting back with regenerative projects that challenge the obdurate, science-based inequalities, or formations, that underlie *neoextractivismo*. Allied with sympathetic lawyers, RSL has challenged the new GM seed laws in court, using prior laws that protect the culture and territory of Afro-Colombian communities to maintain their rights to plant non-GM seeds. RSL is also cultivating "seed schools" and other programs for sharing cultural and scientific knowledge about their own seeds, plants, and practices to help ensure that these do not disappear. And, they are forming relationships with farming communities in other Latin American countries, such as Argentina and Mexico, where similar assaults on traditional agriculture and ecological knowledge are underway (Arancibia 2013; Kinchy 2012).

The conflicts roiling Columbian agriculture have several features that mark it as a scientific inequality formation. First, it contains competing projects, each with science at the center. *Neoextractivismo* is the dominant project, entrenched and anchored by new biotechnologies. The other subordinate project is RSL and its allies who are mobilized for social change, in part through an emergent new science of agroecology. Second, although the two projects have developed different strategies, technologies, and definitions of legitimate science, they are co-evolving in a particular socio-historical context, defined broadly by globalization and the neoliberal reforms that have transformed the international political economy over roughly the past fifty years. This transformation has included an intensification of market-oriented, industry-driven life science and biotech research and the increasing "scientization" of agricultural policy (Kinchy 2012). Third and relatedly, the struggle over seeds in Columbia is not unique to that country. Similar struggles – over seeds, land tenure and land use, and pesticide application, have risen across the global South, in Latin America, Africa, and Asia. The case also raises new questions about the organization and dynamics of scientific inequality formation, involving problems of organizational scale (e.g. how does local knowledge travel "upstream" into national and international domains of science and policy?), scientific authority (e.g. what kinds of knowledge make local farming practices "scientifically" legible to different types of experts?), politics (e.g. what are the key institutions conferring power on different social actors in a given conflict?), and values (e.g. do new communication technologies make science easier to share and access but also harder to control?) We will explore these and related questions in the chapters to come.

Prescriptions for Practice

Even in our current era of "post-truth" politics and consonant "crisis of expertise" (Eyal 2019), science continues to enjoy a special type of social authority tied to the provision of specialized knowledge. It is wrapped up in social mythologies relating to objectivity and value-freedom, economic efficiency and prosperity, the spread of democracy, and quasi-magical powers to problem-solve and techno-fix (Sarewitz 1996). Though often conducted in places and contexts closed off from public view, science continues to be celebrated as a common good. For such reasons as these, scientific inequality formation can be difficult to study; doing so often requires some methodological fine-tuning to bring our analytical sensibilities into proper focus. To that end, we offer a few general prescriptions for carrying our framework and key concepts into practice. Here are four to get us started, and we will spotlight these in the chapters to come to illustrate why we think they are valuable guides for political sociological investigation of science.

1st Prescription for Practice: Science happens in out-of-the-way places. Find and study the spaces where scientific inequality projects are made, altered, transformed, or destroyed.

Science emerges through projects that take place across a wide range of domains, fields, and settings. Lots of STS research bears this out. Different laboratories communicate regularly with one another and also take cues from the "world of commerce" (Kleinman 2003). Science is a common feature of statecraft, and scientists play instrumental roles as unelected policymakers (Jasanoff 1990). Public interest organizations and other organizations run by scientists, such as the Union of Concerned Scientists or Science for the People, provide cultural spaces where politics explicitly intermingle with science (Moore 2008; Schmalzer, Chard, and Botelho 2019). Black, Indigenous, Latinx, and other social groups contest science-based government standards for allowable industrial toxins, challenging the long history of the government's encouragement of profit at the expense of the health of these communities (e.g. TallBear 2013). Because such interactional configurations influence (and often limit) the kinds of questions scientists can legitimately pursue – and thus shape scientific authority – we should focus our analytical attention on the domains, fields, and out-of-the-way, often local and unremarkable settings, where scientific inequality projects are forged, contested, submerged, or implemented.

2nd Prescription for Practice: Study scientific failure and derailment (as well as successes).

STS is chock full of scientific success stories. Our bookshelves are lined with scholarly analyses of prize-winning scientists, research programs that flourish, hot new discoveries, amazing inventions, lucrative patents, and riveting, history-altering facts. Science also is littered with experiments that fail, ideas that fizzle, programs that die, and research bandwagons that run off the rails. We would do well here to recall David Pellow's (2000) assertion that "inequalities are relationships" and not only follow the leaders but the losers as well. Scientific inequality formation is often most apparent at moments when efforts are blocked, re-routed, or shut down. As a result, it is as important to illuminate the power dynamics of these situations – to know when, why, and how science fails – as to seek out additional success stories (Bloor 1991).

3rd Prescription for Practice: To understand how scientific inequality formations change, pay attention to forces working to resist change.

Common wisdom holds that science is an arena of unrivaled creativity and dynamism. STS has played a non-trivial role in reinforcing this view (see 2nd Prescription for Practice). Yet science is an inherently conservative institution, one whose epistemic cultures and systems of governance strongly favor stability and order. It is a world where professors seem never to retire, textbooks are turned out in double-digit editions, and where research findings can take decades or more to harden into fact. Thus, studying scientific inequality formation means entering an already-made world of obdurate structures. In such a world, it is important to study not only how social actors challenge the status quo, as social movement scholars well know, but also how the formation is resilient to organized challenges.

4th Prescription for Practice: Treat scientists and non-scientists like the multi-dimensional actors they are.

There is a strong tendency in STS to depict scientists and engineers as one-dimensional experts focused single-mindedly on professional advancement to the exclusion of other goals, interests, and values. They are thought only rarely to have much to do with political or social issues of the day; the cost, some say, of pursuing "the scientific life" (Shapin 2008; see also Hermanowicz 1998). The perception is false. Scientists are people, too, whose cultural and political lives intersect with their research activities in many ways, some overtly and explicitly, others less so. If we sometimes treat scientists as uniformly unaware of the ways that their work reinforces inequality, injustice, and oppression, the opposite holds for non-scientists: when social movements challenge scientists' authority and power, opponents – including scientists – will often dismiss their concerns as scientifically illiterate, based more on emotions than facts. This perception is also false. Non-scientists reliably use reason and logic to identify problems and posit solutions, just as scientists have a long history of engaging in political activities. It is important to acknowledge both realities and seek out the complexity in actors' motivations for action or inaction (Ottinger 2013a).

For others interested in studying scientific inequality formation, these four prescriptions of practice can offer strategic guidance in where to look for interesting cases, what kinds of cases are worth studying, what kinds of actors are change-makers, and what kinds of forces are operating against change. We have used them to guide our own decisions about what to include, what to leave out, and how to organize the rest of the book.

Looking Ahead

There is much more to say about scientific inequality formation and we use the three chapters that follow to flesh out additional details and further develop our argument. We want to understand their historical development and identify some of the processes and mechanisms that reinforce, alter, or destroy them. Beyond the specifics of particular cases, we want to learn something about broader and longer-term dynamics and path dependencies, about how cases or projects are connected and how those connections can create chains of change, or alternatively, walls of resistance. We also want to begin mapping the contours of contemporary scientific inequality formation and the physical, social, and symbolic spaces these structural forces occupy.

Each of the next three chapters describes different types of scientific inequality projects, marked by a politics of profit-making (Chapter 2), a politics of absence (Chapter 3), and a politics of resistance (Chapter 4). We mean discussion of these three processes to be illustrative and suggestive, not exhaustive or conclusive. We have tended to focus more on research topics that we know best, so our selection of topics and cases shows clear biases toward our own research interests and professional networks. Where possible, we have also aimed for comparative breadth, seeking out different types of cases, the better to convince readers that scientific inequality formation is a useful model for understanding how and why science allocates values in ways that consistently create or reinforce social disparities.

Chapter 2 considers the centuries-old relationship between science and profit-making. Our argument is that scientific inequality is both a product and a source of this relationship and our goal in the chapter is to show how science is conditioned by and reinforces inequality specifically in the pursuit of profit. The historical reach of the chapter is long, including cases from the sixteenth and eighteenth centuries and with a brief detour back to the fourteenth. We chose this approach to convey the importance that our theorization of scientific inequality formation gives to long-term processes of path dependency and also to show how capitalism has been deeply entwined with science – even proto-science – from the beginning. The chapter is mostly situated in the contemporary "neoliberal era," however, which begins in the 1970s and remains a dominant force in science today. In the context of an increasingly market-oriented posture to science policy and knowledge production, we locate a central dynamic of scientific inequality formation in the tension between two main projects. A dominant project that casts science as the engine driving technological innovation and industrial development, paired against a subordinate project mobilizing to bring scientific and public attention to the "manufactured risks" that science and engineering produce (Beck 1992). Case studies illustrate how an embedded logic of profit-making shapes research problem choice, knowledge production, technology development, and regulatory action in ways that both legitimate and fortify scientific inequality formation.

Chapter 3 takes a similarly broad historical perspective on scientific absences as a form of epistemic inequality. Across a range of cases spanning 300 years, we describe different ways that people, research, and knowledge goes missing in science. Our thesis here is that absence-making is not an anomalous feature or bug, but foundational to the system. We argue further that scientific inequality formation *requires* the systematic production and articulation of absence and so it is incumbent upon researchers to get into the habit of asking "absent-minded" questions about scientific business as usual. Doing so allows the often-invisible infrastructures of science inequality formation to come more clearly into view.

Chapter 4 builds on the previous chapters in developing an analysis of resistance to the forces of inequality operating within and emanating from science. Here, we assess science's potential as a political force for social and scientific change – a topic that has occupied sociologists and historians of science for decades. The rise of objectivity as a dominant scientific discourse in the early twentieth century (Chapter 3) operated as a powerful ideological impediment to more progressive or radical forms of scientific activism. Yet more recently, science's increasing entanglement with the logics of markets and profit-making (Chapter 2) undermines scientific claims to objectivity, creating legitimacy problems for scientific institutions and opportunities for deeper substantive reform within science inequality formation. Where science activism was once broadly considered antithetical to science's legitimacy and social authority, collective political mobilization among scientists and science-adjacent professional fields is now increasingly viewed as necessary for maintaining (or, for some, regaining) its social authority.

We end the book not with a set of tidy conclusions, but with a provocation set against the backdrop of compounding ecological crisis, economic fragility, and the global rise of ethno-nationalist political movements roiling the world today. Will science respond in ways that deepen existing inequalities or refashion itself in a way that centers values of human and ecological well-being, justice and community? What can and should that science look like? Lest STS continue contributing to the problem of scientific inequality formation, we believe researchers in our field are morally obliged to seek answers to these questions.

2

Profitable Knowledge

In 1571, alchemist Philipp Sömmering arrived at the north German court of Julius, Duke of Braunschweig-Wolfenbüttel. On the promise of producing a tincture for turning silver into gold, Philipp secured a formal contract from the Duke for himself and a retinue of fellow alchemists. The Duke also offered Sömmering several gifts, including a horse and some English cloth as well as lodging, a laboratory converted from a stable, and several research assistants to help with the work. The work, however, did not go as planned. Treated with suspicion in the Duke's court, the alchemists soon ran afoul of several powerful courtiers, including Julius' wife. Accused of committing fraud, but later also accused of adultery, murder, sorcery, attempted poisoning, and theft, Sömmering and several of those with him were imprisoned, interrogated, tortured, tried, convicted and, in 1575, brutally executed as *Betrügers* – fraudulent alchemists.

As interpreted by science historian Tara Nummedal (2007), Sömmering's punishment as a convicted fraudulent alchemist illustrates the historical construction of a legal-cultural boundary functioning to preserve social belief in Sömmering's opposite: the *true* alchemist who practiced *real* alchemy and *could* transmute base metals into precious ones.[1] If the role of alchemist was not yet fully articulated, alchemy nevertheless represented a legitimate and potentially valuable subject of investigation in those days, with Isaac Newton and Robert Boyle (about whom, more in Chapter 3) counted among its more illustrious European practitioners. Sömmering was the outlier here, for, as recent scholarship has also shown, alchemy occupied a central place in the surge of cultural interest in empirical knowledge associated with the Scientific Revolution and was a "driving force behind the emergence of laboratories, debates about the power of human technology and the boundary between art and nature, matter theory, and even specific ideas like gravity" (Nummedal 2007: 8). His was brilliant proto-scientific derailment (see 2nd Prescription for Practice), exemplifying itinerant knowledge production in out-of-the-way places, like converted stables (see 1st Prescription for Practice).

Nummedal's study focuses not on Enlightenment superstars like Newton and Boyle, but on far more typical, if less studied, "entrepreneurial alchemists," whose livelihoods were supported by the political and financial elites of the Holy Roman Empire and whose practices, importantly for us, were often entwined with industry (specifically mining), state power, and money. As she draws the connections:

> [T]he princes and wealthy investors who supported alchemical work in this period clearly saw more commonalities than differences between alchemy and mining; in fact, these patrons clearly thought about alchemy as an extension of their long-standing interest in mining technology. Patrons hired alchemists and mine experts to address the same kinds of technical problems . . ., and patrons frequently responded to alchemical proposals with the same kind of investor mentality that framed their response to mining proposals. This connection between entrepreneurial alchemy and mining would have important consequences for the early modern practice of alchemy, as alchemists were expected to produce not merely ideas, but also increased profits. (86)

This chapter considers the centuries-old relationship between knowledge making and profit-making. With alchemy, that relationship pre-figured the rise of modern science. We aim to show that scientific inequality formation today remains both a product and source of this relationship.

The word *profit* is usually understood in economic terms, as the money one is left with after accounting for all other expenses. But the term's original meanings – *progress*, from Old French, and *advance*, *benefit*, from Middle English – speak to a more capacious set of ideas that intuitively align with modern notions of science as a means to human progress, intellectual advancement, and broad societal benefit, as we noted in our opening chapter. That science may instead represent a strategy for securing financial profit seems to contradict the common notion that Science (with a capital S) stands (or should stand) beyond the reach of crass human economic or political interest. This contradiction – between an idealized notion of pure, objective, or value-free science and its "never pure" practical embedding in markets, politics, and culture (Shapin 2010) – troubles the title we chose for this chapter and fuels our analysis of scientific inequality formation throughout the book.

Our analysis of profitable knowledge relies on the term "profit-making" rather than "capitalism" or the even more abstract "economy." Like capitalism, profit-making is inherently exploitative; in the process, someone or something is inevitably losing revenue, value, worth, or well-being, relative to someone else (Fleurbaey 2014). Indeed, profit-making is a necessary feature of capitalism, working historically as a mechanism for accumulating and concentrating wealth and power and thus for deepening all kinds of inequality. As we will show, this includes inequities within the domains of science and technology. Yet, unlike capitalism, which is a vast, complex, and dynamic system, profit-making is a relatively straightforward idea that not only predates capitalism historically but also anticipates a certain set of practical activities and goals. It is something most of us have experienced in some way in our own lives, as Sömmering experienced in his own tragically shortened one, and so it is tangible and relatable in ways that a complexly abstract global system like capitalism is not.

We have organized the chapter in rough chronological fashion, the better to illustrate how science inequality formation has both fed on and nourished profitable knowledge across the centuries. The bulk of our discussion is situated in the contemporary "neoliberal era," which begins in the 1970s and remains a dominant force in science today. (STS also emerged in the 1970s and we will have more to say about neoliberal strains of STS later in the chapter.) A central dynamic of scientific inequality formation in this period lies in the tension between two main projects – a dominant project that casts science as the engine driving technological innovation and industrial development, paired against a subordinate project mobilizing to bring scientific and public attention to the "manufactured risks" that science and engineering can often produce (Beck 1992). This contemporary dynamic is not new, but a continuation and multiplication of the ways that profit-making and science have intertwined historically, as we describe in the next two sections.

A Long and Complicated Marriage

Beyond practices specific to alchemy, economic ideas planted the seeds of modern scientific thought. Writing on the origins of science, historian Joel Kaye (1998: 16) has shown that the rapid "monetization" of European society during the fourteenth century – involving the spread of markets and towns, acceleration of agricultural and craft production, innovations in commercial enterprises and techniques, and more generally, the popularization of money as a medium of exchange – had enormous economic, political, social and *scholastic* impacts. With the rise of this new market order, which placed explicit cultural value on practices of measurement, calculation and quantification, the old Aristotelian model of nature as fixed and absolute slowly gave way. In its place, Kaye writes:

> Scholastic natural philosophers began to create a new model of nature, one that could comprehend the order and logic of the marketplace – dynamic, self-equalizing, relativistic, probabilistic, and geometrical – a nature bound together and constructed by lines in constant expansion and contraction. It was within this new model of nature that science emerged. (14)

And as it emerged, scientific ideas and products – including new ways of conceptualizing, measuring and predicting profit and loss – became inextricably bound to the rise and expansion of a new economic order called capitalism.

Writing on economic developments during roughly the same time-period, sociologists Raj Patel and Jason Moore (2017: 24) put a similar point more bluntly. They write, "if profit was to govern life, a significant intellectual state shift had to occur: a conceptual split between Nature and Society" involving "a transformation in how some humans understood, and acted upon, nature as a whole." In their view, the writings of natural philosophers like Francis Bacon (1561–1626) and René Descartes (1596–1650) were instrumental in formalizing philosophical distinctions between nature and society. Their categorization of the world into two overarching and unequal domains "cheapened" the world by setting humans apart from the non-human world and then elevating the intrinsic value of one category ("Society") above the other ("Nature"). The shift involved a "massive exclusion" within early modern notions of society by relegating most women, Indigenous and other colonized people, and non-Europeans to the lesser-valued "Nature" side of the equation, further cheapening all of these exploited "others" in an iterative process in which "the permanent demands of profit-making require those profits themselves to generate profitable returns" (Patel and Moore 2017: 27).[2] The cultural projects led by Bacon and Descartes were foundational for scientific inequality formation.

As Patel and Moore show, profit-making triggers many types of world-historical change in the unending search for new sources of natural resources, new commodities, new markets, and new forms of labor organization to keep the machinery of economic growth lubricated and running smoothly. Political leaders with their states, financial elites with their banks, and merchants with their cargoes, will go to great lengths to ensure that this happens – building colonial empires, opening new frontiers, waging war, and subjugating human and non-human life and nature. The authors describe the uniquely intensive and rapacious forms of profit-making that gained momentum in the early years of capitalist development and that concentrated wealth at every scale of social organization – in regions and countries and within social groups – and could not have emerged as it did without the ideological and practical support of early modern science. Curiously, Patel and Moore have almost nothing further to say about the mutually constitutive relationship between science and profit-making. Even so, their *longue durée* perspective on capitalism's inexorable cheapening of the world encourages our attention to how this long marriage has matured over the centuries and how, in the process, profit-making has functioned as both effect and cause of science inequality formation.

Let's begin with a pair of illustrations from the eighteenth century involving two technologies – one famous, one infamous –

that violently, unevenly, and indelibly, altered the shape of the modern world.

The Chronometer and the Slave Ship

In 1764, a self-trained clockmaker named John Harrison from Lincolnshire, England, designed and successfully demonstrated the world's first marine chronometer, presenting it in London to the Fellows of the Royal Society and collecting an award for his (not inconsiderable) troubles from a British government board. With chronometers on board ships, sailors could determine where they were located on an East–West line around the globe – their longitude. This was a big deal. For millennia, mariners had read the skies to determine their location along a North–South line, or latitude, by locating the celestial North Pole. It was easy to lose one's way, and many ships did. Harrison's chronometer, mounted aboard the British warship *HMS Deptford* sailing from Portsmouth, England, to Kingstown, Jamaica, in 1761–1762, meant that now ship navigators could read latitude and longitude together. The technology made a global positioning system possible and made sailing less dangerous. The voyage to Jamaica provided initial proof.[3]

Harrison's invention helped transform ocean trade. The ability to geolocate a ship's position at sea now meant that ship owners, captains, and traders could do more business more efficiently and reap greater profits. In the years that followed, more ships began making longer direct voyages, and shipbuilders committed more resources to designing and launching bigger ships, better assured of their safe return to port. Inventors, including other clockmakers, clamored to improve on Harrison's original design, and as these refinements went into mass production they further increased the accuracy and reliability of maritime navigation, accumulating power and wealth for Europe's emerging bourgeoisie in a period that economic historians recognize as the waning decades of the age of mercantilism.

Mercantilism is an early form of capitalist exchange, in which traders secured profits by moving basic goods like coffee, spices, or timber from cheap markets like the colonial port in Kingstown, Jamaica, to more lucrative European markets like those in Portsmouth and Liverpool, often tracing the same route that Harrison's chronometer took on the *HMS Deptford*. Yet, what most histories of Harrison's discovery leave out, was that during the late eighteenth century, Jamaica was home to more than 300,000 enslaved Africans forced to work on colonial plantations using rudimentary technologies without pay or consideration for their care or humanity. There were no prizes awarded for solving the problem of the high death rates that plagued Caribbean plantations and resulted in continuous demographic decline among Jamaica's population of enslaved people (Dunn 2007). There were no government boards incentivizing the development of labor-saving designs for the sharp curved knife or "billhook" that those same workers wielded with brutal effect in the intense heat and humidity of the Caribbean cane fields. Even the slave ship – one of the most important technologies involved in the transatlantic slave trade – serving as prison, transport, mobile trading station, and factory[4] all in one – seems to have infrequently benefited from technological innovation, given that the opportunities for greatest profit derived from the products of enslaved people's labor, not their bodily health and well-being.[5] This logic directed technological innovation elsewhere along the mercantilist commodity chain, accounting for its relative absence in the design, production, and operation of slave ships.

According to maritime historian Marcus Rediker (2007), most slave ships were not originally built as such, but were conversions of schooners, sloops, brigs, or retired war ships. Presumably, in time many of these vessels would have carried chronometers as the technology came into regular use. Specialization as slave ships, when it did occur, tended to borrow from existing technologies: cannon for defense, copper-bottomed hulls to prevent rot, wind sails to channel fresh air belowdecks, and shelving, chains, shackles, and neck irons to warehouse and control the captive population. *Sui generis* technologies, like the "barricado" (wooden barricades raised to prevent revolts above decks) and netting strung beyond the hull (designed to prevent suicides from successfully jumping overboard), were simple and inexpensive investments that prolonged life minimally among the merchants' imprisoned human cargo.

Occasionally, slave traders did contract new ships for the purpose, as Liverpool merchant Joseph Manesty did in 1745. Placing his order with Rhode Island shipbuilder John Bannister, Manesty directed that his new ships be built in "a frugal Suitable manner" and "wou'd have as little money laid out on the Vessels as possible," with "'plain sterns', no quarter windows and with little or no work to be done by joiners in the captain's cabin" (Rediker 2007: 51–52). Rediker also notes that medical doctors rarely accompanied slave ships at sea – another way to keep costs low. Thus, while some of the larger slaving ships were expensive to buy and outfit (running into the hundreds of thousands in current dollars), they nevertheless represented a patchwork of cheap technology used to acquire and supply cheap labor from people whose bodies produced cheap commodities for Europe's growing demand for luxury goods (Patel and Moore 2017).

In this broader context, we can better appreciate the chronometer and the slave ship as sociotechnical elements of a colonial-era scientific inequality formation, fashioned from divergent yet complementary innovation strategies: state-led investments in development of the chronometer inversely mirrored merchants' comparative disinvestment in the slave ship, materializing in two machines that compounded the other's utility. The slave ship supplied unfree labor power to fuel transatlantic trade, while the chronometer further enabled and enriched a colonial production system based on enslavement, transport, and exploitation of that same unfree labor. Those colonial outposts featured prominently in the early expropriation of private property, wealth, and power generated in Europe from the transport of basic commodities – including other human beings – characteristic of mercantilist trade policy.

At the same time, the chronometer also signaled the rise of a different form of economic policy – industrialism – in which mercantilism's reliance on moving basic commodities from cheap markets to expensive ones was increasingly replaced by an economic system organized by the factory-based production of new commodities. Augmented by water-powered automating technologies like looms, gins, and mills, as well as armies of workers, industrialism drove profit-making skyward by creating more commodities, faster and more efficiently, eventually undermining colonial systems of enslavement. The chronometer was one such commodity integral to this larger economic transition.

As historian of science Jim Bennett has noted, "By the late eighteenth century all the essential elements of the marine chronometer were in place; what remained for solving the longitude by this method was for them to be manufactured in numbers and taken to sea" (Bennett 2017a: 87). If mercantilism shaped interest among political and economic elites for the chronometer's development, industrialism provided the material resources, (wage) labor power, and production capacity in English workshops and factories to ensure that Harrison's solution to the longitude problem was a practicable one, small enough, affordable enough, and plentiful enough to supply Europe's growing merchant fleet with a new form of technoscientific power.

In comparing the chronometer and the slave ship, we see two elements of scientific inequality formation at work. The history of the slave ship illustrates how the quest for profits can direct scientific and technological resources away from other domains, such as colonial health, tropical medicine, and plantation agriculture, and toward structural projects that legitimate and perpetuate the brutal dehumanization of enslaved African peoples. Conversely, the history of the marine chronometer illustrates how the same quest for profit concentrated scientific resources into domains of knowledge production that enabled stunning levels of wealth accumulation and other economic opportunities for mercantilists, who could take advantage of the resulting innovations a far distance from the horrors of slavery. And not only Harrison's chronometer, but also Isaac Newton's efforts to aid the engineering of more efficient ship hull design and Robert Hooke's experiments in materials resistance – all funded by the Royal Society – were channeled toward and directly benefited long-distance shipping (Merton 1968) and the Atlantic slave trade. Nothing like this level of technoscientific energy and expertise was directed toward the well-being of multitudes of people enslaved or indentured in British colonies around the world, who extracted, processed, packaged, loaded and unloaded commodities transported by Europe's merchant fleets.

As industrialism caught fire and spread over the next two centuries, advances in engineering, math, physics, chemistry, materials science, and mining and agriculture tied science ever more tightly to profit-making. Coal- and gas-powered machines, as well as advances in lighting, building design, and transportation, allowed firms to scale-up production processes, increase profits, and further concentrate wealth. Social sciences contributed as well. Social statistics and censuses emerged as new technologies of quantification that states wielded in controlling growing urban populations (Emigh, Riley, and Ahmed 2016; Schweber 2006). Innovations such as double-entry bookkeeping helped factory managers track increasingly complex material supply chains, schedules, and contracts, while Frederick Taylor's experiments in "scientific management" tightened shop-floor control over a rapidly growing industrial labor force. These technoscientific advances were not random, but patterned by the questions researchers and funders asked (and didn't ask) and by the institutional support that was or was not made available to them. As a result, the men, women, and children whose lives were increasingly dependent on factory jobs – whether employed on unending production lines or displaced from factory work by some of those same "labor-saving" machines – often saw little benefit, direct or otherwise, from these and other innovations and discoveries (Braverman 1974).

These counter-veiling dynamics may seem to operate independently of one another, but the broader context of scientific inequality formation brings the fundamental relationship between science and profit-making into sharper relief. Because science is a social institution built from cumulative practices, with new ideas and technologies developing from prior knowledge, epistemic spaces of intellectual advantage and disadvantage tend to deepen or intensify over time. In this way, science generates path dependencies that follow routes of existing investments and pools of expertise.

In part, these routinized patterns are a consequence of the puzzle-solving logic of scientific inquiry, as Thomas Kuhn (1962) described it, and of the disciplinary specialization of knowledge and expertise. But, in addition, such patterns or path dependencies are also a function – and a feature – of profit-making. Science persists as a bedrock modern institution because, to paraphrase Patel and Moore (2017: 27) quoted earlier, "the permanent demands of profit-making" require profits derived from science to be reinvested in scientific infrastructure, expertise, and practices to ensure the continued generation of profitable returns. The result is dynamic, yet durable, scientific inequality formation, one transformed but not destroyed by the cataclysm of a U.S. civil war (1861–1865) that abolished slavery, but not systemic racism. Our working hypothesis at this point is that the epistemic and social inequalities characteristic of scientific inequality formation will deepen or intensify until changing social, political, and economic conditions substantively alter the distribution of technoscientific resources and power away from profit-making. Our attention for the remainder of Chapter 2 turns to contemporary history, beginning around 1970. This is when most economic geographers, historians, and sociologists point to the rise of a new kind of "knowledge economy" involving a pivotal intensification of profit-making logic into (and through) science and science policy. The term most scholars use to describe the present era, encompassing a dizzying array of economic, cultural, and societal changes – from the globalization of trade, to the branding of personal identities, to the commodification of the water we drink and air we breathe – is neoliberalism.

Neoliberalism and Scientific Inequality Formation

What is neoliberalism?

First articulated in the 1940s by a small group of economists (Mirowski and Plehwe 2009), neoliberalism began as an economic theory that challenged then-dominant policies of social welfare based on the state's provision of public goods, such as education, health care, and a clean environment. Although these and other public goods were never evenly distributed, they were based on the principle that societies thrive when public goods and protections are widely available. When and where these conditions hold, communities tend to be healthier, have higher levels of workforce participation, suffer from less crime, and enjoy more stable families. However, public goods do come with certain costs: they can cut into profits, and higher tax rates are often required to pay for them.

Neoliberalists took a different view of the state's role in society. Rather than relying on government to protect and provide public goods, they believed that government should play a far more limited role, focused mainly on promoting free market trade in part by *removing restrictions that limited profit-making*. Under neoliberal theory, individual behaviors and choices operating in the marketplace unrestrained by government regulations would generate private forms of benefits and protections that would substitute for government-mandated social welfare. This strategy requires fewer taxes and expands opportunities for profit-making. It also retracts social supports for people and groups and pushes an ethic of self-reliance through individual responsibility and entrepreneurship as the preferred route to social and economic well-being.

Early on, neoliberal economic theory found a receptive audience among leaders in Chile, the United Kingdom, the United States, and, to a lesser extent, Germany. By the early 1980s, as national political leaders around the world began to adopt neoliberal principles, international governance systems, such as the World Bank and the International Monetary Fund, began adopting them too (Chorev 2018). Today, neoliberal ideas thoroughly infuse modern societies, characterized by policies aimed at "shrinking state provision of social safety nets and democratic spaces . . .; privileging of free trade and private property; privatization of public goods and ecological and communal spaces; de- and re-regulation," as well as the increasing "financialization" of national and international political economy (Malin and Kallman 2022: 55).

Neoliberalism's core tenets have not only spread around the world, but have mutated and metastasized into new discourses, identities, occupations, and practices that are simultaneously uniform in their global ubiquity and particular in their hyper-local instantiations. For instance, while social media "influencers" may reference a new occupational identity shared by millions, there is only one Kim Kardashian – a neoliberal entrepreneur par excellence. In this way, neoliberal ideology transforms citizens into consumers and valorizes the societal status of individual entrepreneurs responsible for creating and capitalizing on economic opportunities of their own making. As David Hess (2013: 187) writes: "in a world in which long-term employment is precarious, government welfare floors are declining, organizations crave innovation, retirement funds are individually managed, and hedge-fund managers rule the world, everyone's life becomes a story of entrepreneurship . . ."

The consequences of neoliberalism for science have been similarly profound. Under this late-twentieth-century economic order, knowledge itself is treated explicitly as a commodity to be privatized, bought and sold, positioning science squarely inside markets, where the logic of profit-making dominates all other activities and assessments of worth and value, while limiting public support for "collective goods," such as education, social welfare, and environmental protection. As many have argued (e.g. Malin and Kallman 2022), privatization of knowledge contributes to the erosion of community life, of public participation in decisions about technoscience, and ideals of personhood and privacy that are more pronounced for groups that are already marginalized.

In the neoliberal era, the tentacle-like extension and intensification of profit-making throughout modern scientific culture has produced far-reaching changes that have major implications for science inequality formation at different scales of organization. For example, at the macro-level many nation-states now pursue technoscientific innovation as a leading strategy for economic development (Block 2008). At the meso-level, once provisionally non-market organizations such as schools, hospitals, government agencies, and universities are pursuing agendas for profit-oriented research and investments (Slaughter and Rhoades 2009). And, at the micro-level, for many practitioners, science has become an entrepreneurial pursuit as much as a disciplined vocation (Kleinman 2003). In these ways and more, contemporary science has assumed an historically unprecedented commercial value in society – a means to start a business, attract investors, create markets, sell products and services, and ultimately, to turn a profit. And for businesses, science and scientists have become important vectors of economic power. In short, neoliberalism has intensified scientific inequality formation while – perhaps – also creating new means and mechanisms for transforming it (a question we explore in Chapter 4). To get a better understanding of science inequality formation under neoliberalism, we turn next to the early work of pioneering environmental sociologist, Allan Schnaiberg.

An engine? Or a brake?

Writing in the United States in the late 1970s, as government policies guided by neoliberal economic theory were just getting underway, Schnaiberg (1977) published a "social structural analysis" of science and engineering that helps us frame the broader problem that neoliberalism poses for scientific inequality formation. Schnaiberg's analysis begins from a broad distinction between what he called the "technological-production sciences" and the "environmental-social impact sciences" (1977: 501). Production sciences refer to fields such as physics, chemistry, molecular biology, computer science, math, and engineering, whose research agendas and problem foci generally support industrial expansion and economic growth. They do so by producing the basic knowledge lying behind innovations in chemical synthesis, genetic engineering, microchip electronics, particle dynamics, and power generation, for example. By contrast, impact sciences study the social and environmental consequences of industrial production, as represented by fields such as wildlife biology, ecology, environmental toxicology, risk analysis, or even sociology – fields that study the causes and consequences of socioenvironmental problems that tend to follow from industrial development, such as biodiversity loss, air and water pollution, climate change, and environmental injustice.

Building from the premise that environmental protection is locked in dialectical relationship with economic growth, Schnaiberg argued that the scientific division of labor structures the distribution of resources, communication patterns, disciplinary and professional reputations, and social authority of experts and fields in ways that, on balance, privilege production sciences over impact sciences. Any number of comparisons can illustrate this idea: Physics is a higher prestige discipline than ecology. Biochemistry attracts higher levels of funding than conservation biology. Engineers play a more central role in shaping environmental policy and land use decisions than sociologists. Top economics departments are far less likely to hire in heterodox research areas like ecological economics than in more orthodox areas of the discipline. Historically, the directors of the National Science Foundation and the White House Office of Science and Technology Policy (OSTP) – key public figures in setting national research and science policy agendas, respectively – have represented classic STEM fields: math, physics, computer science, engineering, and, less often, life-science fields like microbiology or genetics.[6] Virtually all of them have been men. Sociologist of science Dr. Alondra Nelson's appointment in 2021 by the Biden Administration as OSTP *Acting* Director is a rare double exception – a female social scientist – to the patriarchal/ production science rule.[7]

In the aggregate, a dialectical imbalance between a dominant production and subordinate impact science "obscure[s] the negative consequences of high-technology production" for the environment and society (Schnaiberg 1977: 509), as problems created by production science grow in scale and complexity, outpacing remedial solutions derived from impact science, which fall further and further behind. Conflict is endemic to this uneven dynamic, with "maverick" or "activist" impact scientists working for change from within, while government oversight agencies such as the now long-defunct U.S. Office of Technology Assessment (1974–1995) and social and environmental movements press for change from outside. Regardless, because such inequities are institutionalized in funding decisions, research agendas, career trajectories, and professional norms, Schnaiberg concluded that resistance from the impact sciences is unlikely to result in lasting structural changes. As he put it, "If . . . the growth of knowledge is the most important dimension of the [alternative technology] revolution, it is clear why the bulk of such knowledge has reinforced the [power of] existing high-technology producers, and has led to little melioration of environmental consequences of such production" (517).

As readers will by now have likely surmised, Schnaiberg's typology, if a bit crude, summarizes succinctly the two main discursive projects we suggest are constitutive of contemporary scientific inequality formation. The dominant project, represented by production science, is commonly described today as an engine fueling market-oriented innovations and profit-making. The subordinate project, represented by impact science, operates more like a set of over-worn brakes, continually but often gingerly, pushing against the status quo to bring scientific and public attention to the social and environmental problems – including social inequalities and environmental injustices – that production science inevitably creates alongside all its shiny objects, from pills and smartphones to particle accelerators and Nobel Prizes (Beck 1992).

Moreover, four decades on, the framework continues to evince scholarly relevance. A study by Schnaiberg's former Ph.D. student Kenneth Gould (2015: 145) describes disparities in funding and communication patterns shaping the emergent field of nanotechnology, where "nano-impact science has been dwarfed by nano-production science," even as leading environmental groups continue to raise public awareness of the uncertainties associated with nanotechnology risk and demand increased government funding for nano-impact science (Gould 2015: 149). Similar kinds of disparities are routinely reported in the context of environmental higher education, where most environmental studies programs "suffer from limited resources or unequal standing relative to the traditional disciplines" (Vincent et al. 2016: 418). Elsewhere, we see how the push–pull of discursive projects becomes institutionalized in the development of agroecology and associated community "seed schools" as forms of scientific resistance to *neoextractivismo* in Latin America described in Chapter 1. We also see the dynamic play out in in conflicts over climate policy, where activist scientists from groups like Scientist Rebellion are increasingly embracing civil disobedience as strategies for slowing the fossil-fueled climate emergency (Frickel and Tormos-Aponte 2023).

Even as the framework remains broadly relevant, the conditions shaping scientific inequality formation that Schnaiberg described four decades ago are themselves changing, growing more acute and more extensive under neoliberalism. For example, from 1950 to 1980 U.S. expenditures for research and development (R&D) were shared more or less equally between the federal government and private industry (National Science Board 2022b). By 2020, industry was footing the bill for 72 percent of national R&D. The private sector also now plays a more outsized role in shaping national and disciplinary research agendas than in the past, in part by funding a steadily increasing proportion of R&D conducted in academic and government settings. As a result, industry is now also the largest "performer" of R&D activities (i.e. the actual doing of science and technology), accounting for about 75 percent of total, compared to 12 percent and 9 percent performed by universities and the federal government, respectively (National Science Board 2022b).

The trend is not unique to the United States. For example, a National Science Foundation report on global science notes that:

> Within most of the top R&D-performing countries, the business sector funds the most R&D – 60% or more in 2018. In each of the leading Asian countries – Japan, China, and South Korea – the business sector accounted for more than 75% of R&D funding. (National Science Board 2022b)

Parsed by type of activity, private sector labs in the U.S. are producing nearly all research geared toward experimental development of new technologies (90%) and more than half of all applied research (58%). These statistics are not all that surprising given that tech innovation and market development serves industry's direct economic interests. More surprising is that industry's share in the performance of basic research – historically, the centerpiece of university research portfolios – is significant and growing, up from 21 percent in 2010 to 32 percent in 2019. More broadly, according to the International Science Council:

> The private sector's share of global science and innovation is growing, and is now estimated to represent approximately 70 per cent of global expenditure on science. At the same time, publicly funded researchers are increasingly encouraged to form partnerships with the private sector and to undertake research that will support private priorities, while the commercialization of academic research is increasingly regarded by government as a priority for universities. (https://council.science/actionplan/science-private-sector/)

Statistical trends offer a bird's-eye view of decadal changes to scientific inequality formation, as corporate interests in profit-making dynamically shape science funding, research agendas, and knowledge generation, as well as constraining societal access to increasing shares of basic, yet legally proprietary, knowledge.

Such shifts underline a generally underappreciated reality of scientific inequality formation: private firms hold proprietary rights to a rapidly increasing share of the scientific and technical knowledge produced globally, including large and increasing shares of basic scientific knowledge. As science privatizes, the logic of profit-making increasingly shapes, but also obscures, decisions governing what knowledge gets made, who makes it, and who benefits. This dynamic is playing out across many different fields of technoscience, from AI to defense research to pharmaceuticals. It is what makes accessing science conducted in privately controlled spaces (like corporate laboratories) resistant to structural changes that diminish scientific inequality formation is so important (1st and 3rd Prescriptions for Practice).

Prefiguring our focus in Chapter 3, on the absences created by scientific production, privatization simultaneously shapes what knowledge is left "undone" (Hess 2016), who is excluded from knowledge-making, and who is harmed as a consequence of the absence of knowledge. Yet, aggregate statistics can mask more subtle, but broadly impactful, cultural changes shaping scientific inequality formation as neoliberalism erodes institutional boundaries and loosens cultural mores that traditionally have distinguished – although never decoupled – academic from commercial science.

Academic Science Becomes Big (and Little) Business

In the United States, the structural conditions for neoliberalizing university research came in 1980 when Congress passed the Patent and Trademark Act Amendments. Commonly referred to as the Baye-Dohl Act, so-named for the two U.S. Senators who sponsored the bill, the Act incentivized profit-oriented research by allowing universities and individual faculty members to patent, own, and commercialize innovations generated from government grants. For the first time, U.S. universities could legally compete with one another for profits creating a new kind of "knowledge economy." Since then, various other federal and state policies have further incentivized universities to reorganize academic planning and budgeting, hiring, curricular and program development, teaching, research and careers, in ways that better align with the neoliberal order – one that interprets higher education increasingly through the lens of market growth and profit-making (e.g. Etzkowitz 2008; Lave, Mirowski, and Randalls 2010; Rudy et al. 2007; Slaughter and Rhoades 2009).

The list of such changes is long and varied, but one key feature has involved a shift from traditional liberal arts education to inspire lifelong learning, training in scientific methods and analytical critique, and engaged citizenship, to professional training that positions graduates for post-graduate employment in specific job markets. In the U.S., this shift was evident in the 1980s in the creation or growth of professional schools to advance careers in business, health care, hospitality, architecture, and engineering, among other sectors (Brint and Karabel 1991), in the sharp growth of new community colleges in the 1990s and 2000s, and in the current boom in online education (Education Dynamics 2023).

Another key feature of neoliberalism's increasing influence on academic science is the adoption of market-oriented university funding models that shift the costs of higher education downward, from societies to individual students and their families. A marker of this shift is the global trend toward privatization of higher education (OECD 2021). Around the world, but particularly in Latin America and Asia, where privatization has been most intense (Saforcada and Socolovsky 2019), once-free university education now increasingly comes with a price tag. In the U.S., public universities have adopted a number of related strategies to counter lost revenue from cuts in federal and state spending on higher education. Most notably, they have raised tuition rates. At 4-year public colleges, tuition expenses rose 45 percent between 2000 and 2022 (NCES 2022). Many have also increased student enrollments, sometimes by lowering admissions criteria or by investing in campus amenities like swimming pools, fitness centers, and luxury dorms, to outcompete other universities for high-value tuition dollars. The explosion of terminal master's level degree programs at U.S. research universities reflects a complementary strategy to raise new forms of revenue from graduate education programs. Another fiscal strategy that many larger research universities have adopted is creating innovation centers, investing in existing departments that encourage technology development, and encouraging faculty to patent discoveries and form commercial start-ups. Some scholars worry that a market-oriented funding model will have corrosive long-term impacts on the types of technoscience that universities produce, with decisions to grow, shrink, or dismantle existing departments based on market signals like student job opportunities or patenting potential rather than quality of learning or the social value of knowledge and ideas (Rudy et al. 2007; Nickolai et al. 2012).

Yet another sign that in the era of neoliberalism, higher education is valued increasingly as more of a private service than a public good, is that universities are adopting governance models that shift control of budgets, programming, and hiring decisions away from academic departments and into the hands of upper administration. Often accompanying centralized governance is a parallel shift toward an "audit culture" (Strathern 2000) in which faculty and departments are assessed through regular performance reviews, common features of the corporate world. The reviews themselves are now based less on qualitative assessments of one's substantive ideas, teaching acumen, and ability to mentor students, than on quantitative metrics such as the number and value of external grants one receives, publication rates, "impact factor" scores, course enrollments, and student satisfaction, as measured by course evaluation surveys. Quantitative metrics are now favored by university administrators and governing boards because they facilitate "objective" comparisons among faculty and departments. These metrics are folded into branding strategies and translated for public consumption onto university websites to attract potential students.

The strategy is "neoliberal" because it treats the campus as a marketplace, where faculty resources, tenure and promotion, salary increases, time off, teaching assignments, and service loads are distributed competitively rather than strictly meritocratically (in the case of tenure), or even democratically (in the case of service assignments). As the stratification of academic life has intensified, departmental autonomy has shrunk, working conditions for many university employees have deteriorated, and intellectual relationships – among administrators and faculty and among faculty and students – are now more transactional than ever. Organizing access to resources under a neoliberal logic tends to emphasize growing revenue streams, rather than increasing student access to quality education.

Yet, it is not just that the larger-scale political economy of higher education and R&D under neoliberalism is shifting technoscientific power away from universities and toward the private sector, as Allan Schnaiberg earlier predicted. More subtle and diverse cultural processes and mechanisms are also at work in creating and reinforcing scientific inequality formation from inside. For example, scholars have identified an "ethic of entrepreneurialism" (Moore et al. 2011) or a "code of commerce" (Kleinman 2010), where universities, departments, labs, and faculty are rewarded or punished, in various ways, for taking risks and pursuing new projects that promise to translate knowledge into marketable products and profits, or not. These and other cultural transformations are changing what kinds of scientific knowledge get produced, and how, as ethnographic research by sociologist of science Steve Hoffman has shown.

Hoffman investigated decision-making processes inside two artificial intelligence laboratories to understand how neoliberalist ideas are shaping strategies for managing ambiguity in scientific research. One lab he studied relies on large federal grants and is oriented toward academic research; the other pursues investor funding and industry contracts and is oriented toward commercial technologies. The cultural differences he found are striking.

Where the academically oriented lab built its research program around disciplinary theories of intelligence using broadly shared experimentalist approaches, lab members expressed far less concern about whether their research had market applications. This lab culture produced knowledge that was deeply theoretical and methodologically rigorous, but also incremental, "myopic," and largely untethered from broader market and social considerations.

In contrast, the commercially oriented lab conducted research by "avoiding [theoretical concerns and debate], relaxing and blurring epistemological standards, and elevating concerns with user design to the heart of their science" (727–728). The ethos in this lab enabled researchers to be "responsive to intermittent funding opportunities, to quickly move in and out of new problem spaces as they arose, and to produce commercially viable systems that were not limited by prior ontological commitments" (728). Hoffman describes the knowledge generated by this lab as "bold and exciting science" that comes at a cost of also being "disconnected, thin, and deeply inflected by the capital concerns of industry or the latest fads of popular consumer markets" and which "choke off serious interrogation of theoretical assumptions or the identification of unifying themes" (728; see also Hoffman 2021). Where an entrepreneurial ethos takes hold, especially in academic research settings, Hoffman (2021: 570) concludes, "we might worry that sustained engagements with the thorniest of scientific problems can wither away as the champions of academic capitalism tell stories about how quick and clever they have become."

Hoffman's emphasis in the last statement on the importance of entrepreneurial stories that work to legitimate the market orientation of academic science conforms to our argument in Chapter 1 about the role of dominant and subordinate discourses in the theory of scientific inequality formation we are developing. His research also reminds us that to understand inequality in science we cannot limit our analysis to macro-level structures. In different contexts and using different strategies, universities, departments, and laboratories are also chasing technoscientific power, acting more and more like private firms competing in the marketplace and embedding market logics into academic decision-making and curricular and research practices (Vallas and Kleinman 2008). The organizational components of science are mechanisms for market expansion, just as markets are catalysts for scientific research and teaching. In other words, neoliberalism has created new conditions for the intensification and multiplication of scientific inequality formation across different scales of social organization. Increasingly, even Schnaiberg's "impact science" is no longer immune. As Gould's (2015) study of nano-tech shows, today "green" impact sciences – including green chemistry and materials research, alternative energy and energy storage technologies, and geo-engineering – are big business, increasingly indistinguishable, discursively, organizationally, and economically, from production science.

STS's Cultural Aversion to Inequality

The eroding forces of neoliberalism have shaped the contours of STS in material and conceptual ways as well, with real consequences for how the field understands and studies scientific and other forms of inequality. As Philip Egert (2013: 13) observes, "STS departments are not immune to the neoliberal currents sweeping universities, such as continually increasing tuition rates in which students become consumers and debtors, an increasing emphasis on research partnerships with industry, and the hiring of a growing community of part-time transient adjuncts." These currents were not always detrimental, as the economic value of STS expertise have risen in some sectors of the academy. In the U.K. and U.S., for example, some STS researchers rode the wave of neoliberalism into newly created positions in university business schools (as noted in Wooglar et al. 2009); others have turned their analytic attention to the study of financial markets and the knowledge practices of stock traders (e.g. Preda 2023). In addition to changing conditions of STS labor brought on by neoliberalization of the academy, the conceptual history of the field also bears the mark of the neoliberal economic order, as recounted by Steve Fuller (2000) and David Hess (2013).[8]

STS emerged in the 1970s as an intellectual movement anchored mainly in Northern Europe. From the beginning, proponents took pains to distinguish the new field from North American sociology of science, led by Robert Merton, as well as from old-school Marxist critiques of science and technology. Both traditions, in different ways, were political in the sense that they expressed particular ideologies and developed structural arguments about the social organization and impact of science and technology in society.

Marxists claimed that the history of science, and especially the history of technology, could be read as directed by the dominant interests of the capitalist political economy and that the class interests of scientists and engineers in different disciplines and industries helped explain the social organization of science and the direction of technology development (Mannheim 1936). Structural forms of inequality are central to Marxist analyses of science, often maligned by critics as "deterministic." Less overtly, Mertonian scholars read science through the ideological lens of liberal democracy. This meant treating the relative autonomy of science from the state and capitalism as a starting point for the study of culture and values shaping scientific institutions. The main problem for this group of sociologists centered on understanding what Merton (1973) called the "normative structure of science" as that value system shaped the meritocratic assessment of scientific ideas, the distribution of scientific resources and status, and equality of opportunity among scientists (Hess 2013). Like the Marxists, a major focus of Merton and colleagues (e.g. Cole and Cole 1973) was inequality, but now framed as a problem of social stratification. Detractors cast aspersions on the Mertonian project as a "functionalist" apologia to science.

In contrast to both of these older traditions, early proponents of STS declared its central project to be the "sociology of scientific knowledge" (SSK), arguing that the content of scientific knowledge, heretofore treated by historians, sociologists, and philosophers of science as off-limits analytically, was not only amenable to social and historical analysis, but a necessary target if one's goal was to understand why the knowledge claims of scientists are judged by peers to be true or false. SSK eschewed Merton's institutional analysis, which left explanations of the content of scientific knowledge intentionally unexplained, and rejected the crude structuralist arguments of Marxist political economy. Instead, laboratories (Latour and Woolgar 1979), as centers of knowledge production, and technical disputes among experts (Collins 1974), as fora for the negotiation of scientists' knowledge claims, were held out as invaluable sites of social investigation.

But, at least early on, as David Hess (2013) points out, SSK still made room for structural explanations amenable to the study of inequality. They did so by using the concept of "interests" – now broadened beyond material class interests to encompass a wide variety of experts' social and cognitive commitments – as a vehicle for structural explanations of scientific knowledge and resulting inequalities. A classic study in this vein by Donald MacKenzie (1978) concerned a dispute among early twentieth-century statisticians about how best to measure statistical associ-ation. Drawing evidence from published and unpublished work of the principal antagonists, MacKenzie identified differences in Karl Pearson's (1857–1936) and George Yule's (1871–1951) "cognitive interests," or the specific goals that propelled their understanding of the mathematical problem to be solved.

Consonant with our 4th Prescription for Practice to treat scientists are multidimensional actors, MacKenzie also found differences in the antagonists' "social interests" derived from their membership in different sectors of British society. Pearson belonged to a rising middle class of professionals that espoused strong commitments to eugenics theory and social policy, in part because the new professionals exemplified the tenets of "posi-tive eugenics" in affording a privileged social status to intellec-tually superior individuals (see, e.g., Adami 1921). Although the younger of the two, Yule represented the old elite, then very much in decline (as fans of the popular BBC series *Downton Abbey* may appreciate), and presumably uninfluenced by the eugenics movement, which were hostile to the idea of aristocratic rule. Thus, MacKenzie concludes, "Differing social interests can be seen as entering indirectly, through the 'mediation' of eugen-ics, into the development of statistical theory in Britain" show-ing that "'hard sciences'. . . should not be excluded a priori from analysis in terms of social interests" (71–72).

In the early 1980s, consonant with the celebrated embrace of neoliberalism by the Reagan and Thatcher governments, "inter-ests explanation" like the one MacKenzie used, came under attack in STS (Woolgar 1981a). Playing out largely in the pages of *Social Studies of Science*, the debate revolved around the question of whether scientists' social interests provided the nec-essary empirical foundations for explaining the content of scien-tific knowledge.[9] As ethnographer Steve Woolgar (1981a: 371) argued, in one of the debate's more forceful critiques:

Interest-work is . . . constitutive of scientific practice. But . . . we have
as yet little appreciation for the way this kind of work is done by
scientists. It is at best inappropriate, therefore, to use interests as a
resource at the expense of investigating how they are accomplished.
The construction and use of interests is an aspect of scientific activity
which demands treatment as a phenomenon in its own right.

From Woolgar's perspective, the unseen phenomena that
MacKenzie was calling interests – often appearing as latent or
implied "goals," "purposes" or "commitments" – were the very
things that needed explaining; as such, STS should not rest its
accounts of science on them. A more empirically justifiable goal
for the new field, Woolgar argued, would be for scholars to produce detailed descriptions of the scientific practices by recording *what scientists actually do*, including how they communicate
with one another in making, defending or contesting one another's knowledge claims. Pursuing this approach, the field would
accumulate more empirically nuanced understanding of *how* science works by studying scientists' day-to-day activities.

Woolgar's arguments, among others, had impact. As the dust
of debate settled, interest theory largely disappeared from STS
and with its erasure, structural explanations of science – and the
field's long-held attention to scientific forms of inequality – came
to occupy a decidedly subordinate analytical status. The field's
main objective across most of the ensuing decades[10] has been
to show "how agents actively perform or construct their world
rather than how their worlds shape them" (Hess 2013: 182).

This historical dynamic reflects the push–pull of scientific
inequality formation inside STS, pitting "a dominant network
of agency-based frameworks and subordinate networks that
retained a greater interest in structural explanation and extra field
influence on the scientific field" (Hess 2013: 186). Driving that
tension are variously branded versions of constructivism[11] and
their tendency to rely on "empirically vague and politically non-
committal concepts like 'hybridity' and 'co-production' of science and society" (Hoffman 2017: 728). Such concepts, and the
approaches embedding them, eschew a direct analysis of power
and inequality. Instead, they promote a general model of science
that substitutes structural explanations for narrower descriptive
analyses of entrepreneurial actors, the micro-contexts in which
researchers and technicians are said to operate, and the largely

unfettered translation of their ideas into regulatory and commercial products. None more so than the "actor-network" theory of technoscience developed by Michel Callon and Bruno Latour (1947–2022), a framework reportedly favored by organization scholars and business and management researchers (Woolgar et al. 2009: 8, 23, and note 13).

As philosopher of science Steve Fuller argues, however, actor-network theory's popularity may rest on its "affinity for the metaphysics of capitalism" with its emphasis on commodification, productivity, efficiency, and treatment of "humans as cogs in the wheels of a machine and machines as the natural producers of value" (Fuller, 2000: 374; for the full critique, see 365–378; see also Hess 2013). Our contention is that science inequality formation operates to some extent within all academic fields, including in STS where its impacts have been realized most acutely in the field's intellectual move away from structural explanations and investigations of scientific and social inequality. For now, we end this chapter on profitable knowledge in a more grounded way, by considering how regulatory agencies deploy science in response to the economic threats posed by contaminated soil.

Conclusion

"Building No. 2" is a four-story, 66,000 square foot brick, cast iron and concrete edifice located on the banks of the Woonasquatucket River in downtown Providence, Rhode Island, and former headquarters of the Brown & Sharpe Manufacturing Company. Famous by 1900 as the world's largest precision tool factory, for nearly a century (from 1872 to 1964) dozens of company engineers, and thousands of machinists and other laborers worked at the 33-acre industrial complex designing and building machines that built other machines – lathes, mills, grinders, and gauges – providing the material foundations for industrial-scale production and profitability. Today, the refurbished building is a showcase of twenty-first-century eco-conscious urban redevelopment, replete with "LED lighting, the latest in heating and cooling technologies and solar panels."[12] It houses the state's premier environmental agency, the Rhode Island Department of Environmental Management (RIDEM) and, interspersed among scores of employee cubicles in the cavernous 3rd-floor Office of

Land Revitalization and Sustainable Materials Management, nearly 3,000 environmental site investigation reports that catalogue the state's cumulative knowledge of soil contamination. Each report summarizes a particular site's land-use history, describes the current conditions of the lot, presents results from chemical analysis of soil and groundwater samples, outlines site remediation plans and lists restrictions on future site reuse. The subject of one of these reports, identified as file SR-28-0495 F and dated from March 18, 2002, is Building No. 2, indicating that RIDEM headquarters is also a RIDEM hazardous waste site.[13] Building No. 2, its history, its current political–administrative occupants, and most importantly the site investigation reports compiled there tell a story – or set of stories, really – of scientific inequality formation, state power and profitable knowledge (Frickel 2023). Although most site investigations are managed by RIDEM, they represent the collective work of dozens of private entities who are key players in New England's burgeoning environmental services industry (on the industry generally, see Macfarlane 2019). Some of these companies are smaller, local outfits, but others are large internationally scoped companies that employ thousands. For example, the site investigation report describing environmental conditions for Building No. 2 was created by the Louis Berger Group, a U.S.-based "full-service engineering, architecture, planning, environmental, program and construction management, and economic development firm" that was listed as a top revenue-generating design firm in 2015 and was acquired by a Canadian company, WSP Global, in 2018.[14] The same site investigation report referenced additional contributions from an architectural firm, a drilling company, and two analytic laboratories and represent the private production of public knowledge.

Federal environmental statutes embed RIDEM site investigations squarely within the context of local real estate transactions. The economic logic of property frames contamination as a problem of individual "sites," that are to be judged as risky or safe for development and use through a graduated process of discovery as soil and groundwater is collected, samples are tested, and results are interpreted. These scientific practices, combined with the dynamics of property values and real estate markets, shape the generation of site-specific regulatory knowledge packaged in the reports that RIDEM produces. Because they are tied to real estate transactions, state-sponsored information about soil contamination tends to concentrate in neighborhoods where real estate markets are dynamic, but accumulates more slowly in areas that are economically depressed or in neighborhoods blighted by poverty, industrial decline, and institutional neglect. In this way, entrenched social and environmental inequality is accompanied – and likely reinforced – by a type of state-sanctioned epistemic inequality, in which some neighborhoods consistently generate more regulatory attention and knowledge than others in ways that map onto the ebb and flow of property values in commercial real estate markets. Importantly though, while federal environmental statutes tell us how regulatory science is supposed to work, legal abstractions tell us little about whether policies governing site investigations are enforced and how they are actually implemented. When real estate agents, lawyers, buyers and sellers are gathered in a room to ink a property sale, their collective interest in not finding evidence of contamination – that would immediately gum up the works of a property sale – may operate as a form of what sociologist Linsey McGoey (2019) calls "strategic ignorance" that grants plausible deniability to those involved in the transactions.

Thus, like the alchemists who introduced this chapter, the regulatory scientists, engineers, and technicians who work for RIDEM are performing a new sort of alchemy, using regulatory science to turn bad soil into good. Like the material and ideological support that Duke Julius provided Philipp Sömmering, regulatory knowledge also derives much of its epistemic legitimacy from the legal and political authority of the state. And, like those innovators of early modern scientific practice, the data collection, testing, mapping and evaluative analyses deployed in site investigation reports today, in Rhode Island and across the United States, are deeply entwined with the interests of powerful economic actors – in particular, those industries connected directly or indirectly to real estate and development. The epistemic inequalities this scientific inequality formation produces are marked by profitable knowledge but also, in some neighborhoods more than others, by its absence – the topic of our next chapter.

3

Absent-Minded Science

In the days before Hurricane Harvey slammed into the Texas Gulf Coast in late summer of 2017, the Texas Commission on Environmental Quality (TCEQ) shut down its stationary air monitoring equipment. The stated reason was to protect the monitors from the storm, which would soon pummel some parts of the Houston metro area with more than 60 inches of rain and 150 mile per hour winds.[1] On the agency's advice, Texas Governor Greg Abbott next suspended air pollution emissions rules and waived environmental reporting requirements; these would remain suspended for eight months.[2] Asked to shut down production three days prior to Harvey's landfall, many local industrial facilities waited until the storm moved onshore to shutter, a delay that resulted in the unplanned and potentially illegal release of at least 8.3 million pounds of air pollutants (Phillips 2018), but now without the standard slate of TEQ air monitoring stations to document declining air quality conditions. Harvey battered area chemical plants, oil and gas refineries, pipelines, and storage tanks (Zhang 2017). Floodwaters inundated several of the area's dozen or so highly contaminated "Superfund" sites, causing sporadic fires and explosions that sent additional pollutants billowing in the air and washing millions of gallons of fuel- and chemical-laced wastewater into nearby neighborhoods (Dearden and Biesecker 2017).

Once Harvey had moved inland and weakened, damage assessment to the heavily industrialized region got underway.

Scientists from NASA's Earth Science Division offered to fly a DC-8 jet equipped with "the most precise and comprehensive airborne air quality lab on the planet" over Houston to test whether air pollution levels had changed appreciably since the disaster (Rust and Sahagun 2019). Following a flurry of emails between the agencies, TCEQ officials refused NASA's offer, with the Texas Director of Toxicology stating "we don't think your data would be useful" and that local air monitoring being conducted by state officials was "sufficient" (ibid.). Officials at the U.S. Environmental Protection Agency then endorsed TCEQ's refusal. Facing fierce pushback from both agencies, NASA's Earth Science Division Director reluctantly cancelled the data collection flight. The plane and its sophisticated monitoring equipment remained parked in the airfield hangar.

Did damage to industrial infrastructure from Hurricane Harvey reduce air quality in Houston and raise health risks for area residents? We may never know for sure, although one computer modeling study found that pollutant aerosols released from chemical plants during the storm likely acted as "cloud condensation nuclei" in quantities substantial enough to *double* the amount of rainfall in neighborhoods close to industrial facilities (Pan et al. 2020). More rain means more flooding, and another study found that 84 percent of flooded parks in metro Houston were potentially impacted by nearby toxic exposure sources such as Superfund sites or landfills (Karaye et al. 2019), quite possibly compounding further the health risks to residents living near industrial areas.

However, direct evidence from atmospheric testing that could have provided more definitive knowledge about air quality dynamics and hurricane-induced air pollutants' impacts on human and non-human health are missing. They are missing because the NASA scientists who possessed the know-how, resources, and good will to generate the data were excluded from the decision-making process, their expertise rejected by Texas agency leaders as unnecessary. It would also render the state's own data largely unfalsifiable.

The story of NASA's grounded data-collection flight suggests that both industry, which valued production over safety, and the state, which valued their air monitors more than the air itself, were both making decisions that advanced the logic of profit-making, the topic of the previous chapter. But if we want to understand scientific inequality formation as something more than the simple outcome of capital accumulation and economic interest, we would do well to investigate how economic and political power operates by excluding certain types of knowledge and knowers from decisions impacting scientific research and policy. And, to do that, we need to pay close attention to absence, our main topic for this chapter. It's a common adage that knowledge is power. But the *absence* of knowledge, equipment, and people with relevant skills and perspectives can articulate as power too, including marginalized groups and others with on-the-ground knowledge, as illustrated above. We develop and refine this idea in the following pages.

Our central thesis is that scientific inequality formation depends on – indeed, it requires – systemic productions and articulations of absence. Implicit in this thesis is the idea that absences in science do not arise "naturally," or of their own accord. Instead, academic and regulatory science generate absences of various sorts, intentionally or unintentionally, on a regular, ongoing basis. That is, absences are not anomalies in science, but institutional outcomes, baked-in to scientific business as usual. We will have more to say about our conceptualization of absence in the next section.

A key goal for the chapter is to identify some of the social mechanisms or regularized processes that generate absences in science and, more broadly, shape science inequality formation. As our introductory vignette just illustrated, absences are created by denying access to technologies that would allow the production of new knowledge. Absences are also created when existing knowledge is sequestered or otherwise prevented from circulating. Knowledge practices can be erased intentionally by those in power, such as industrialists, politicians, or leaders of academic disciplines. Knowledge can also be sequestered by those with little power, as a protective strategy. Knowledge practices can fade away through casual neglect or indifference, for example, when a crisis diverts researchers' attention or when an investigator retires and their work is set aside by others who have different research priorities. People can be left out too, intentionally or unintentionally, and thus excluded from specific labs, academic programs, or entire domains of research. And people who made important impacts as knowledge producers can be forgotten or actively erased from history, along with their scientific and social contributions. We are interested in how different ways of producing absence pattern scientific knowledge and practice and thus contribute to the formation of more durable scientific inequalities.

In the following pages, we introduce a series of cases that are designed to bring the production of absence into sharper focus. The cases are diverse, ranging across centuries and continents and disciplinary domains. Together they provide a menu of different processes or mechanisms that generate absence. These include concrete examples of knowledge suppression, exclusion of social groups, historical erasure and the social structure of insurgent intellectual networks. When regularized as part of institutional practice, these and other mechanisms structure scientific inequality formation. Before we get to our empirical cases, though, we need to better situate absence conceptually, methodologically, and historically.

Situating Absence

According to some, absences are integral to the preservation of social order and general societal well-being, something to be celebrated for their positive impacts on society. This is the argument that Princeton sociologists Wilbert Moore and Melvin Tumin (1949) made about ignorance – an absence of knowledge or "not knowing" – in an earlier sociological treatment of the subject. They saw ignorance as "both inescapable and an intrinsic element in social organization generally" (788, note 4). Illustrating the post-WWII dominance of structural functionalism[3] in North American sociological theory, Moore and Tumin's short essay identified five ways that ignorance serves a positive "social function," for example, by maintaining social hierarchy (because continuity of social structures depend upon differential access to knowledge) or by reinforcing traditional values (because tradition depends on the absence of legitimate alternatives). Their theory did not address the role of absence in science, however, nor did it highlight the politics of knowledge that comes with absence.

Nevertheless, while structural functionalist thinking fell out of favor among sociologists decades ago, Columbia University neurobiologist Stuart Firestein (2012) more recently imported functionalist logic into a book about science written for popular audiences. In it, Firestein argues that ignorance does not preserve the status quo in science, so much as operate as a fundamental driver of scientific research and innovation. According to Firestein, science begins in ignorance, at the limits of what is known, and is a product of science too, insofar as new discoveries inevitably raise new questions that redefine the limits of knowledge. Scientists' collective search for new kinds of ignorance – not new knowledge – "functions," in Moore and Tumin's terms, to impel science forward.

Our view is different still. As our opening vignette of the grounded NASA data collection flight illustrates, absences in science do not just happen. Instead, and as many before us have noted, absences are *sui generis* social productions (Smithson 1989; Proctor and Schiebinger 2008). They are the result of distinct, socially and historically situated processes that require their own social explanations, and they reflect arrangements of power, interests, and identities. Institutionally prescribed mechanisms that generate and maintain absences – be they absences of knowledge, technology access, people or practices – are not exceptional or trivial. Neither are they random. Rather, they are fundamental to the operating efficiency of modern science and a key to its cultural and political authority – an authority that science derives in part from scientific inequality formation. These are big claims, so we want to be clear about what we mean and why.

The absences we care about are those that operate in patterned ways across domains to shape research networks and agendas, open or close down funding streams, and invisibly structure relationships between universities, governments, and firms. The authority that science claims and preserves within these processes are based not only on the promise of a better world or on the promise of patentable discoveries and marketable technologies. Scientific authority also rests on the credentials of experts and institutions, a system that necessarily excludes other voices, knowledge and practices and is therefore generative of social relations that construct, reinforce, and challenge scientific inequality formation. We see power and the politics of absence at work here. We think studying and thinking carefully about the politics of absence in science – centering on non-production or non-circulation of knowledge – can tell us a lot about how and why science works better for some than for others.

But to do so, we need different theories and different methods for accessing the unknown. Let's start with the basic observation that theories of knowledge will not tell us much about the absence of knowledge. Because absences are *social productions*, it is not enough simply to read the nonproduction of scientific knowledge inversely from cases of knowledge production, as if absence was something static and uniform, like a photographic negative. Instead, absences of knowledge are often recognized through complex processes of social recognition that typically involve the mobilization of concern by social movements and other civil society actors who draw critical attention to "undone science" (Hess 2016; Frickel et al. 2010) or to the "invisiblization" of knowledge (Henry 2017; Kuchinskya 2014). In the Houston example, an investigation by two *Los Angeles Times* reporters raised public awareness and led to a congressional hearing and increased public pressure for transparency among the environmental and health agencies assessing post-disaster risk.

Similarly, we should not read institutional failure in science inversely from theories intended to explain scientific success. We should not assume, for example, that the theory of scientific merit (e.g. Hagstrom 1965) can sufficiently explain why so few women Ph.D.s are employed as computer scientists. The theory of scientific merit is a theory organized by ideas about the social nature of recognition in the scientific community and seeks to explain why some people build successful careers in science. It does not explain why most people who build successful careers in science are men (Smith-Doerr, Alegria, and Sacco 2017; Alegria and Branch 2015; Fox, Bunker Whittington, and Linková 2017). To understand why computer science employs relatively few women Ph.D.s, we need a different theory: one focused on mechanisms of exclusion that systematically deprive certain groups of knowers from gaining access to powerful social networks and material resources in science. If we understand more generally that the absence of a diverse range of people from the highest ranks of science is unjust in itself and limits the perspectives and insights on which scientific knowledge is built, it follows that the study of absence will require its own conceptual toolkit.

Another important challenge we encounter when studying absence in science is that, by definition, the object of study itself is missing. How does one study something that is not there? As sociologist Linsey McGoey points out in *The Unknowers* (2019), successful efforts to hide damaging information for strategic purposes are nearly impossible to track systematically; we only know about non-disclosures when they are exposed to public view. Identifying missing data or methods can be tricky, especially when absences are hiding in plain sight, so to speak, in scientists' daily work routines or disciplinary cultures (Croissant 2014; Frickel 2014a; Knorr Cetina 1999). The absences of relevant groups from knowledge making are easier to trace, though there are formidable barriers to redressing them. In Chapter 4 of this book we will discuss the role of social movements and other collectivities in building social recognition of missing data and knowledge and in mobilizing public resources to "get undone science done" (Ottinger 2017: 560).

Above all, absence study requires nourishing a sensibility toward the unknown and the missing. Toward that end, we have found that it helps to be absent minded. Not in the figurative sense rendered by the gendered caricature of the (always male) absent-minded professor,[4] but in the literal sense and without cynicism, as in "mindful of absence." For us, being absent-minded means treating what is missing (people, data, knowledge, technology) as data that should be analyzed and explained rather than dismissed as trivial or explained away as a source of error or unaccounted bias. It also means asking questions that focus, somewhat unconventionally, on the non-production of knowledge and on the "regimes of imperceptibility" that, as M. Murphy (2006: 9) writes, frame "the history of how things come not to exist." Why were Houston residents deprived of cutting-edge science and technology that NASA scientists might have produced in the wake of Hurricane Harvey? How do such exclusions limit scientists' already tenuous capacity to assess environmental risk across Houston and other storm-damaged cities? Absent-minded researchers should be asking this type of question, although this has not always been the case.

Social Studies of Absence

Social scientists have been studying absences, off and on, for a long time. Key thinkers from the nineteenth and early twentieth centuries, including Karl Marx, Sigmund Freud, Max Weber, and Georg Simmel, developed theories about why some knowledge goes missing or is never made to begin with (Gross 2007; Harding 2006). Even so, it took researchers much longer to recognize the topic's social and scientific importance.

In 1987, Robert K. Merton argued that scientific fields take stock of what the field knows and what questions remain unanswered. All fields do this, he suggested, but some more than others, and it results in what Merton called variable patterns of "specified ignorance" or, the social recognition of missing knowledge. Anticipating Firestein's insider account of scientific ignorance by several decades, Merton argued that specified ignorance is functional to scientific disciplines because accounting for missing knowledge "lay[s] the foundation for still more knowledge" as he demonstrated with an analysis of serial attempts to identify missing knowledge in sociological theories of deviance.[5]

A few years later, in an essay that drew attention to mundane technologies that indelibly shape everyday social life, Bruno Latour (1992) asked "where are the missing masses?" The missing masses he had in mind were things like seatbelts and speed bumps, which encourage the observance of traffic safety laws even when drivers are not being that observant, and automatic door closers that act on their own to regulate indoor air temperatures even when people forget to close the door behind them. These artifacts are not missing in an empirical sense; indeed, they are ubiquitous. But at the time, they were missing from sociologists' analyses of social groups and explanations of human behavior. Latour argued that excluding these and many other common modern technologies limited sociologists' ability to fully grasp the complex sociotechnical materiality of contemporary society and appreciate the extent to which social action is inescapably conditioned by technology.

Both essays represent attempts by sociologists of science and technology to grapple with the problem of absences. Despite beginning from very different assumptions about the nature of scientific knowledge and the institutions that support it, neither author used absence as a conceptual tool for studying science per se. Instead, Merton and Latour take up the problem of absence introspectively, using it to think about the ways in which social scientists, and sociologists in particular, produce knowledge. Both conclude that we can produce more and better social knowledge by taking (different kinds of) absences into account. We agree, but their justifications are far too narrow.

In the past two decades, scholarly interest in absence – and in kindred concepts such as ignorance, agnotology, erasure, refusal, forbidden knowledge, invisibility, and undone science – has surged (see Gross and McGoey 2015; Frickel 2014b; Kempner et al. 2011; Kuchinskya 2014). With this newer work – much of it originating from communities of feminist STS and critical gender, race, Indigeneity, and sexualities scholarship (e.g. Mills 2008; Sullivan and Tuana 2007; Tuana 2004; Tuck and Yang 2014) – a missing world is emerging from the shadows, a recovered world that promises to substantially redraw our maps of science and revamp our theories of how, where – and for whom – science works. We draw from some of this newer research in the sections that follow. We hope to persuade readers that absences, although often invisible or unrecognized, are nonetheless central features of scientific inequality formation and count among the core elements of its social and epistemological architecture. The three sets of cases summarized below illustrate a few of the specific mechanisms through which absence shapes scientific practices, materials, ideas, careers, and disciplines in ways that harden scientific inequality formation but sometimes also create opportunities for collective contestation and reform, empowerment, and survival.

Flower Power

We begin with historian Londa Schiebinger's (2004; 2008) fascinating account of the peacock flower (*Poinciana pulcherrima*) in colonial and European science during the eighteenth century. The plant, its medicinal properties, and its political value to the project of empire, links the rise of botany in England to the disappearance of cultural practices in its West Indian colonies. The story illustrates how absence emerges through the suppression of Indigenous and female knowledge and the erasure of knowers even as it nourishes state power, global trade, and the discipline of Western science as distinctly masculinist pursuits.

The peacock flower is a plant whose medicinal properties as an abortifacient were commonly recognized by Indigenous and African women in the West Indies. In the context of colonialism and the plantation agriculture system organized around enslaved labor – including childbearing as an economic imperative for agricultural production – this knowledge and the practice of abortion itself were potent tools of enslaved women's political resistance. Keeping this knowledge hidden from colonial doctors and scientists would have likely also served as a source of agency and as a survival strategy.[6] Of course, the plant's ability to terminate unwanted pregnancies would have the same effects among European women as it had among enslaved and indentured women in the West Indian colonies. Yet, these knowledge practices failed to reach Europe, even as the peacock flower itself arrived safely via merchant ships and came to flourish in Old World botanical gardens where it gained favor as an ornamental. Schiebinger (2008: 151) asks why, in its nautical passage from the West Indies to Europe was this powerful "technology of resistance . . . ignored, left to languish increasingly in rumor and innuendo"? Her answer is complicated, but compelling.

In the colonial West Indies, men outnumbered women by substantial margins and keeping men alive and healthy enough to work, whether they were free, enslaved, indentured, or inscripted, was a priority for colonial administrative elites. As a result, colonial medical practices paid scant attention to the medical care of women, especially non-European women. As important is it was for the planter class to keep male workers, troops, and managers alive and able-bodied, it was also imperative to deny women the power to stop their pregnancies, especially after European countries began outlawing the importation of enslaved Africans. Herbal abortifacients such as the peacock flower worked directly against the economic and political interests of the mercantilist state. So, while colonial physicians cultivated, studied, and reported the peacock flower's medicinal properties and uses, they rarely used the plant to induce abortions – although African midwives, for whom the absence of formal knowledge could operate in this instance as a form of power, surely did (see Schiebinger 2017).[7]

Unlike their West Indian counterparts, European physicians did not conduct medical safety testing, publish results of clinical studies, or record the plant's effectiveness in terminating pregnancies in compendia of officially recognized medicines or *Pharmacopoeia* (Schiebinger 2004: 178) – even though colonial reports were readily accessible and the plant itself was thriving in private and public gardens across Europe. Professional and disciplinary interests likely played some role in this "nontransfer

of knowledge" (ibid.: 153). As the medical field professional-
ized, male physicians began to replace midwives and with the
occupational shift, abortifacients gradually lost their mooring in
the domains of women's health and childbirth; when abortion
was required, male physicians preferred physical methods and
surgical instruments over herbal remedies. As Schiebinger (186)
explains:

> The rise of "scientific" medicine and systematic experimentation
> with medical techniques in the late eighteenth century did not ...
> lead to the testing and development of abortifacients [which, in turn]
> did not enter mainstream European drug testing and did not become
> a part of academic medicine or pharmacology as these fields devel-
> oped in the eighteenth and nineteenth centuries.

The explanation that emerges from the historical record is that
physicians considered abortifacients dangerous – and they were,
with powerful pharmacological effects, but also as a hidden
source of female power to control unwanted pregnancies. But
they remained dangerous because, when medical scientists could
have pursued systematic "development and testing of safe and
effective abortive techniques," instead they "chose the road
toward suppression of these knowledges and practices," excluding abortifacients from drug testing, medical study, and classroom lectures (190). In the nineteenth century, as European
states began to criminalize abortion, they also classified abortifacients as poisons and outlawed their use. And so, "as midwives
were run out of the profitable end of their profession (there was
always work for them among the poor), abortifacients gradually
disappeared from mainstream medicine" (193).

More broadly, the case of the peacock flower's suppression by
medical science (and, we must assume, intentional erasure among
midwives who pushed the practice underground) shows how
"funding priorities, global strategies, national policies, structure
of scientific institutions, trade patterns, configuration of technologies all pushed investigation toward certain parts of nature and
away from others" (Schiebinger 2008: 152). This is a powerful set
of ideas that can inform our understanding of scientific inequality
formation. As Schiebinger describes it, absence flows from but
also reinforces the structure of cultural, economic, and political
relationships that institutionalize how and why scientists study
some parts of nature but not others, who does the work and who

benefits from the resulting production or suppression of knowledge. Here, the marginalization of women (as colonial subjects, as enslaved persons, as midwives), the exclusion of abortifacients as topics of medical research, and the subsequent suppression of medical knowledge and practices based on women's needs and experiences reinforced a scientific inequality formation grounded in empire, racism, and misogyny. It did so by deepening women's already unequal decision-making capacity in society, short-circuiting potential benefits the peacock flower's abortifacient properties might have otherwise provided, while simultaneously increasing the actual health risks the lost knowledge had previously mitigated.

More broadly, the case helps make visible the interconnections between inequality within scientific fields (in this case, eighteenth-century botany and medicine) and other societal projects of racial and gender inequality, such as subordination of women and the suppression or outright destruction of Indigenous cultures through settler colonialism and chattel slavery. Later in the chapter we will show how academic disciplines – sociology and anthropology, rather than botany and medicine – operate to reproduce scientific inequality formation. Although Schiebinger is writing about how absence of Indigenous knowledge structured the production of Western scientific roles and disciplines during the eighteenth century, similar structures operate today. Indeed, recent scholarship describes the impoverishment of contemporary Western scientific and regulatory practices resulting from systemic exclusions of Indigenous knowers and knowledge in fields ranging from genomics, where molecular ancestry approaches articulate Indigenous peoples as "vanishing" (TallBear 2013), to Western wildfire ecology and management where fire suppression practices have led to "an erasure of cultural landscape, of particular artifacts, and of the future ability to learn from the ancestors and the land" (Norgaard 2019: 104). Our next case considers how, throughout the nineteenth century and into the twentieth, discourses of scientific objectivity paradoxically came to require the exclusion of scientists themselves.

Objectivity, or the Curious Case of the Missing Scientists

When the peacock flower was crossing the Atlantic during the seventeenth and eighteenth centuries, the legitimacy of scientific knowledge was not free-standing as we tend to think of it today, but rather was explicitly attached to the social authority or credibility of individual scientists. Boyle's Law, for example, describing the inverse relationship between a given quantity of gas and its volume under constant pressure, is named for Robert Boyle (1627–1691), whose published experiments offered the first proof. We might say that European science at that time was distinctly *person*-ified. One's claims to truth rested squarely on a foundation of personal rather than institutional trust and, subsequently, one's position in society. In this particular historical context, the hierarchy of the laboratory mattered a great deal (Shapin 1994). Patrons of the Royal Society and other institutions of learning placed high value on the social positions and cultured skills and graces of "gentlemanly" natural philosophers, many of whom were drawn from the more elite spheres of society. Yet the technicians, whom Boyle paid to assist him and who labored in setting up and conducting experiments and recording the results, remained invisible in this process, lacking names and identities; they were also denied authorship on published papers. As Steven Shapin (1989) describes the moral economy of Boyle's lab:

> Boyle was the *author* because Boyle possessed *authority*. It was he who presided over the scientific work place – indeed, it was his house; it was he who possessed the acknowledged right to set the agenda of work, who could effectively command the skilled labor of others, who could define the boundaries between skill and knowledge. (560)

On the other hand:

> Servitude compromised technicians' political integrity in the community of science and affected their credibility. Who could rely upon the testimony of people who were constrained? Servants might make machines work, but they might not make knowledge. (562)

Indeed, technicians' mistakes as described in Boyle's lab books, "came to constitute an understood moral resource for explaining and excusing experimental failure" (560). This is a clear example of how absence – in this case the manufactured invisibility of laboratory labor – operates within scientific inequality formation to reinforce the authority of (some) knowledge and (some) knowledge makers.

Today, scientific inequality formation renders distinctions between scientist and technician differently than in Boyle's time. Although honorifics like Boyle's Law are still the norm and scientific reputations remain important markers of professional power, modern institutions like double-blind peer review (prevalent in the social sciences) go some way toward decoupling scientific knowledge from close association with particular individuals. Science's social authority is now commonly understood to rest more on the system's capacity to generate *impersonal* knowledge about the world. Thus, trust in knowledge – today signaled on Google Scholar by citation counts, h-indexes, and journal impact factor scores – often outweighs the trust derived from one's familiarity with certain people. We do not mean to suggest that exclusions and erasures of social groups are no longer a problem of contemporary science. They are and they remain central mechanisms for reinforcing scientific inequality formation, as we aim to show. For example, invisible laboratory labor remains a common feature of experimental science today (Bangham, Chacko, and Kaplan 2022; Doing 2004), in some ways making inequality in science harder to see than in Boyle's time.

Yet, compared to Enlightenment-era science, science in the twenty-first century when social movements like #MeToo, Black Lives Matter, Climate Justice, and La Via Campesina have refocused public attention on academic, government, and corporate practices of workplace equality, fairness, diversity, and inclusion, processes of exclusion and erasure tend to come about less overtly, less explicitly, and with less transparency. For example, new research showing strong public trust in algorithms like those used to generate rankings in Google Scholar and which are developed by a coding community that is predominantly white and male, sits uncomfortably against studies of racial and gender bias in machine learning technologies (Benjamin 2019). As told by historians Lorraine Daston and Peter Galison (1992; 2007), the shift from personal to impersonal criteria for judging the worth of science received an important boost in the latter half of the nineteenth century, when scientists first marked their knowledge with a new kind of moral virtue: objectivity.

During this period, these authors write, "'Let nature speak for itself' became the watchword of a new brand of scientific objectivity" marking a distinct shift in the epistemic culture of modern science from one based on relations of personal trust to one based on impersonal, "objective" knowledge (Daston and Galison 1992: 81). Their study documents this shift by examining changes in thousands of images reproduced in scientific atlases published between the eighteenth and early twentieth centuries, as the personal and necessarily subjective hands of skilled illustrators gave way to images produced by machines.

Scientific atlases are picture books, usually over-sized, and designed to showcase how different fields of science "see" the objects of nature they study – for example, human anatomy, plants, fossils, or the solar system. Such atlases are still published today and, in fact, you may have one on your coffee table. Created for specialist and popular audiences alike, in the 1850s these books proliferated and readership soared. "Manifestos for the new brand of scientific objectivity," atlases publicized the social utility and moral worth of scientific knowledge not only through the content of the images, which varied by discipline, but also through the quality of those images and how they were produced (ibid.: 81). A close chronological reading of these images reveals atlas-makers supporting an ever-greater reliance on technology to represent natural objects as unblemished by human interpretation: Hand-drawn illustrations steadily gave way to machine-rendered images using a range of new technologies including lithography, photography, and X-rays. Their embrace of what Daston and Galison call "mechanical objectivity" sought to replace scientists' skillful yet inherently subjective "eye" for the truth with technologically generated images, which ostensibly removed human subjectivity from the process. As camera obscura, lithographs, X-rays, and photographs came to replace the handiwork of artists and illustrators, virtuous science became marked by the absence of individual illustrators and artists in representing the subjects of scientific work. Collectively, these images offered audiences a new vision of science, as a vocation demanding "the honesty and self-restraint required to foreswear judgment, interpretation, and even the testimony of one's own senses" but that also requires at the same time, "the taut concentration required for precise observation and measurement, endlessly repeated around the clock" (83). This new vision of science is one that eschewed the "moody brilliance of the genius" of elite natural philosophers like Robert Boyle in favor of "the plodding reliability of the bourgeois" (ibid.), a decidedly more modern trope typecast for a rapidly industrializing world. By the 1900s, although still evolving, mechanical objectivity had become a core feature of "the professional ethos of science" (Daston and Galison 1992: 122).

A few decades later in 1942, Robert Merton gave objectivity an extended, and distinctly sociological, expression as one of four institutional norms that he saw as preserving the autonomous functioning of scientific institutions, reward structures, and careers. As Merton (1973: 270) described it, objectivity was codified within the scientific "norm of universalism," which holds that, "[t]he acceptance or rejection of claims entering the list of science is not to depend on the personal or social attributes of their protagonists; his [sic] race, nationality, religion, class, and personal qualities are as such irrelevant." Today, this idea persists in calls for "value-free" or "objective" science or sciences free from "bias." And, it lives on in condemnations of "junk science" and in societal reverence for "technical" knowledge, imagined to be free from human influence and above politics.

Objectivity and universalism are kindred ideas that enshrine scientific categories of worthy knowledge and worthy behavior. The two ideas are discursively aligned and mutually reinforcing. Objectivity holds that good science must be impersonal. Universalism holds that scientists' personhood is irrelevant to the production of good science. They are twin pillars in modern scientific epistemology, governing the rules of practice for "how we know what we know" (Knorr Cetina 1999). Each idea has received substantial critical attention from sociologists of science and STS scholars.[8] However, it is the historical construction of objectivity and universalism during the first half of the twentieth century as the epistemological underpinnings of scientific inequality formation that most interests us.

In the framework we are proposing, objectivity and universalism are discursive projects in scientific inequality formation. They operate together as social categories that confer legitimacy and authority on select groups of knowers and types of knowledge and exclude others, even as they profess to advance a system of knowledge production that is increasingly inclusive, transparent, and free of systemic bias. Indeed, one can read the history of objectivity (and universalism) as an idea whose qualities reflect the positions, interests, and emotional detachments of white middle- and upper-class men in the Global North who have always dominated science, in contrast to women and people of color whose subordinate social positions and life experiences prevent them "being objective," that is, from standing apart from their personal experiences. We can also read these discourses as projects to reduce professional scientists to one-dimensional knowers (see 4th Prescription for Practice). As Donna Haraway (1988) and others recognized some time ago, the twin discourses of scientific objectivity and universalism embody a fundamental contradiction. The contradiction is that science claims authority as a universalizing, timeless "view from nowhere," in part by ignoring its own historical "situatedness" in academic, government, and corporate institutions that have always privileged the perspectives of Western, white, male interests and experiences. It is a context-blind discourse dependent on racial, gender, and class-based exclusions.

One consequence of this discursive project is the marginalization of groups whose social positions and ways of knowing place them outside of dominant, conventional categories. Consonant with this argument, sociologist Erin Cech and colleagues (2017: 744) see objectivity and universalism as propping up the "epistemological dominance" of modern science by excluding alternative, non-dominant epistemologies and reproducing systemic disadvantage among historically underrepresented groups in academic research and education. To understand how epistemological dominance works, these researchers conducted in-depth interviews with a sample of Native American college students in science, engineering and health fields who adhere to or revere Native American epistemologies (Cech et al. 2017). In sharp contrast to scientific epistemology, which is understood as singular and universal, Indigenous epistemologies are plural and often place-based. They promote a more open, collaborative relationship with the natural world and embrace complexity and chaos over reductionism and compartmentalization, to better meet the physical, social, and spiritual needs of Indigenous communities (Kolopenuk 2020; TallBear 2013; see also Harding 1998). At issue is how Native American students, who embody non-

Western cultural and historical perspectives, navigate Western science's epistemological "claim[s] to produce universal truths and where the social locations and identities of [scientists] . . . are supposed to matter least" (Cech et al. 2017: 745).

Study participants indicated that, whereas they understood the world to be organized holistically based on native practices and teachings, the science presented to them in college was grounded in reductionism, or seeing small elements as the foundation of larger systems. Students raised to value assisting others in learning as a key part of native culture, confronted course instructors who taught them that competition, not cooperation, was a key to their future success as scientists. Nearly all of the students interviewed for the study hoped to return to a tribal reservation or other native community to assist in addressing technoscientific problems. According to Cech and colleagues (2017: 761), however, earning the credentials needed to qualify for such jobs, "simultaneously delegitimates Indigenous ways of knowing while insisting that Indigenous subjects who attend nontribal colleges adopt dominant scientific ways of knowing before they can 'give back' to those very communities." Instructively for us, they argue that, "these mechanisms of inequality are distinctly epistemological, in that they stem from power dynamics embedded in conflicts between legitimated and marginalized ways of knowing" (762). Although this study is specific to Indigenous Americans in the United States context (see also Kimmerer 2015), epistemological dominance in one form or another is characteristic of scientific inequality formation more generally.

What do these two studies – of the construction of scientific objectivity during the nineteenth and early twentieth centuries and of contemporary Native American college student experiences with STEM education – tell us about absences in science? At a minimum, they provide evidence that science has a missing person problem. Or, more accurately, science has missing *persons problems*, for there are manifold ways that members of social groups and their ideas go missing in science. As we show next, demographic and epistemic absences can reverberate across intellectual networks, structuring them in ways that are profoundly consequential for scientific inequality formation.

Whatever Happened to the
Du Bois–Atlanta School of Sociology?

In the late nineteenth century, a new cognate form of "scientific sociology" arose at Atlanta University in Atlanta, Georgia (Morris 2016). The city and its university were unlikely settings for the birthplace of a new discipline. First, the school was located in the "wrong" part of the country: the historiography of American Sociology places the discipline's origins in the elite private universities of the urban north, in particular, the three storied departments at the University of Chicago, Columbia University, and, a bit later, Harvard University and then Brown University (see Camic 1995). Atlanta University, a small, poorly resourced, state-funded, segregated Black college deep in the American South, was none of these. Second, the first-generation sociologists typically credited with building the discipline – people like Franklin Giddings, Albion Small, William Sumner, W.I. Thomas, and Lester Ward – were accomplished men[9] with strong ties to influential politicians, natural scientists, philanthropists, and university presidents and trustees. W.E.B. Du Bois, who founded the Department of Sociology at Atlanta University in 1897, was also well-connected, better trained than many of these men, and outpublished them all. However, they were white and he was Black. For this, Du Bois and the Black sociologists he trained (women and men) encountered repeated discrimination and marginalization, direct and indirect, working within a racial caste system that governed not only southern society, but the northern academy as well.

Unlikeliest of all, given the group's marginalization by white sociologists, Du Bois and his colleagues nonetheless managed to build an institutional, conceptual, and methodological scaffolding for a new discipline – one that was scientific in its empirical and comparative approach to American racism and reformminded in its aim to overcome race-based oppression and inequality. Anticipating by a century calls for a "public sociology" (Burawoy 2004), these scholars offered up pioneering research on Black lives and Black society that is only now gaining recognition as the "Du Bois–Atlanta School of Sociology" (Morris 2016). The case, rendered in rich detail by sociologist Aldon Morris in *The Scholar Denied* (2016), illustrates both how scientific inequality formation erases individual scholars and scholarly groups as disciplinary knowers and as knowing subjects of disciplinary projects and how intellectual networks can organize and sustain structures of resistance.

The pace and breadth of the work undertaken by the Du Bois–Atlanta School was extraordinary. In the space of just thirteen years, between Du Bois' arrival at Atlanta University in 1897 to his departure in 1910 to work for the NAACP, he and his team created the first sociology department in the south (and one of the first departments in any U.S. region), the first sociological survey and field research laboratory (the Atlanta Sociological Laboratory) and the first annual meeting (the Atlanta Conference). The conference, which predated the American Sociological Society's first national conference by nearly a decade, attracted top academics and progressive reformers from Europe and the U.S., including Harvard anthropologist Franz Boas and Settlement House Movement leader Jane Addams.

The volume and scope of the empirical research emerging from the Du-Bois–Atlanta School was similarly impressive. Du Bois and colleagues pioneered studies in rural sociology, urban sociology, political sociology, sociology of crime and criminal justice, demography, family sociology, cultural sociology, and, of course, the sociology of race and racism. They did so by conducting field and survey research in dozens of rural and urban African American communities across the Southern, Midwest, and Mid-Atlantic states. From this data, they published dozens of original reports and studies and regularly gave academic and public lectures based on their findings. Along the way, they provided on-the-ground training in sociological methods and analysis to dozens of undergraduate and graduate students (Black and white) drawn from around the country, some encouraged by their professors at other universities who had themselves attended earlier meetings of the Atlanta Conference. In this way, Atlanta University's Sociology Department and Laboratory became a hub for an international network of empirically minded social scientists and a proving ground for first- and second-generation Black sociologists.

The Du Bois–Atlanta School was very much a racial project, in Omi and Winant's (1994) sense, aimed at transforming discursive practices in sociology and using new understandings of the impacts of racism on Black lives and communities to redistribute power within the dominant scientific inequality formation. It did so by institutionalizing an alternative sociological space for the study of race and racism and for Black sociologists to carve out successful academic careers. How is it, then, that in most sociology textbooks and historical treatments of the discipline, discussion of sociology's founders excludes all but cursory mention of Du Bois and his School (Sica 2007), or that a white, second-generation sociologist from University of Chicago – Robert Ezra Park – is broadly credited with initiating the sociological study of race? Aldon Morris' answer to these questions highlights the conflict inherent in scientific inequality formation as the organizational efforts of marginalized actors induce fierce opposition from the center. He (181) writes:

> To fully understand why Du Bois's scientific school was marginalized and erased from mainstream sociology, we need a theory of how scientific schools emerge, take root, and become institutionalized that explores the linkages between science and the power structures of the larger society: structures of politics, economics, race, and societal elites. Nothing short of such a focus can account for why Du Bois was denied a broad stage on which to practice his craft in academia.

As Morris explains, compared to his white counterparts in northern elite universities, Du Bois operated with a serious deficit of the types of power that accumulate as "academic capital" (Bourdieu 1973). Atlanta University could not compete with Harvard, Columbia, and Chicago for prestige, resources, or the most promising graduate students. Du Bois struggled to secure the funds he needed to carry out his Laboratory's research projects. Other sociologists could – and did – ignore Du Bois and his team's rapidly growing body of empirical research. He did not command the attention from northern whites as did his ideological opponent, the conservative Black scholar Booker T. Washington and, as a result, Washington could use his own academic capital to marginalize Du Bois. At the height of his School's productivity, Du Bois was not invited to help organize or invited to attend the International Congress of Arts and Sciences in St. Louis in 1904, which gathered together "the leading experts in every branch of science," nor was he asked to present at the International Conference on the Negro in 1912 (Morris 2016: 23, 109). In all, Du Bois's subordinate position in the emerging academic field of sociology – a field structured from the beginning by the racist, Social Darwinian hypothesis that Blacks were intellectually and morally inferior to whites (Morris 2016: 87–88) – made his School's erasure from the field all but inevitable and held back the critical sociological study of race and racism several generations.

Du Bois did succeed in producing a wealth of scientific sociology, but it remained invisible to most of the academic world during his lifetime and far beyond. He did so, Morris argues, by acquiring "liberation capital," where his supply of academic capital fell short, and by harnessing the power of that capital through "insurgent intellectual networks." Liberation capital is "a form of social capital used by oppressed and resource-starved scholars to initiate and sustain the research program of a non-hegemonic scientific school" (Morris 2016: 188). Southern middle-class Black scholars were primed by sociology's liberatory promise to disprove the dominant Black inferiority hypothesis and dismantle the racial structures that held it in place. Du Bois accumulated this capital by building "insurgent intellectual networks" composed of college students, pastors, educators, social activists and others who, because of their race and political commitments were "denied access to elite intellectual networks" and "formally organized intellectual discourse" (ibid.: 193). The insurgent intellectual network would, "donate resources and volunteer labor of activists to develop and validate counterhegemonic ideas; provide previously untrained students and others with needed scholarly tools; create media designed to make this scholarship visible to both scholars and consumers of these ideas; and seize opportunities to challenge and replace dominant paradigms" (ibid.). As Morris stresses, liberation capital and the insurgent network of researchers that drew their power from it can sustain a new scientific school for a time, but they are not sufficient for such a school to flourish and thrive.

The Du Bois–Atlanta School case exemplifies how collective actors build, challenge, and sustain scientific inequality formation. Cultural projects define, tacitly and explicitly, what counts as relevant knowledge. Structural projects organize and authorize scientific education and work. Together, these projects operate in the context of legal and economic arrangements that exclude some groups while leaving others with greater power to influence research agendas and affect knowledge content. As Morris (2016:

193) notes, extant theories of science have difficulty explaining "why great science can go unrecognized by the mainstream because they neglect the interaction between science and political/economic/racial processes." Scientific inequality formation helps us connect the dots to appreciate better the racially complicated history of sociology. It can also help us recognize regularities that connect sociology's complicated history to the history of other disciplines, such as anthropology, where Black American physical anthropologist William Montague Cobb (1904–1990) led a similar project aimed at combating the racist assumptions and practices governing the anthropological sciences at a critical moment of that discipline's institutional development.

The similarities between Cobb and Du Bois are striking. Like Du Bois, Cobb was the first African American to earn a Ph.D. in his field before joining the faculty of a historically Black school, the Howard University Medical School. He was prolific too, with more than 1,100 publications in a wide variety of medical and academic journals, as well as popular journals and magazines. Like Du Bois, Cobb also worked tirelessly for racial justice, first in research that challenged dominant racist assumptions in his discipline. Later, in antiracist politics that informed the building of educational systems and networks for Black Americans, who were legally barred from entry into the best schools in the country and as president of the NAACP from 1976 to 1982. And, like Du Bois, Cobb's important contributions to science were largely ignored, as they are still today. Harvard anthropologist Franz Boas typically gets credit for introducing cultural theories of racial difference into the biology-oriented discipline of early twentieth-century physical anthropology, not Cobb. Yet, according to Rachel J. Watkins (2007: 193), Cobb's perspective on the cultural and historical determinants of race were distinct from Boas' and Cobb deserves independent recognition as "one of the forefathers of the biocultural synthesis."

A major difference between the two Black scholar-activists is that Cobb did not have access to funding that would have supported an ongoing team-based research program like Du Bois'. Instead, he focused on teaching, giving his anatomy class to more than 6,000 Howard medical and dental students (Alic 2022). As a result, he did not cultivate a group of research associates that would form an insurgent intellectual network to build on and extend the research he pioneered, as Du Bois was able to do. He did, however, establish the Laboratory of Anatomy and Physical Anthropology and used the time he could commit to his own research that would supplement and inform his classroom lectures and demonstrations aimed at advancing studies of the scientific characterization of musculo-skeletal variation in Black bodies.

Cobb's racial anatomy project was anchored by an absence in the dominant scientific networks: "the lack of 'Negro materials' in established human skeletal collections" available to physical anthropologists at the time (Watkins 2007: 188). The absence of a representative sample of African American skeletons stymied research that could upend dominant theories asserting Black peoples' physical and mental inferiority. Like Du Bois, Cobb worked to fight the scientific racism then thriving within anthropology with empirical science. "It is my belief," Cobb wrote in 1941 to the Howard University Medical School Dean,

> that physical anthropology can make a significant contribution to our national welfare if it would by giving the people . . . the scientific facts we have about race. In this way, a great blow could be struck at the dominant group's entrenched belief in its racial superiority . . . I do not believe that we can look to others to do this job for us. Nearly every distinguished living American anthropologist, and I know them all now, has private reservations about the intellectual possibilities of the Negro. We cannot expect them to be willing to go very far. (Cobb, n.d.; quoted in Watkins 2007: 188)

Recognizing the need for empirical ammunition in the academic battle against scientific racism, "Cobb assembled a research sample of African American skeletons from cadavera beginning in 1932 until 1969." In addition to skeletal remains, the collection grew to include other physical data such as "anthropometric measurements and limb, hand, and foot X-rays of living populations" as well as social and demographic data that located specific bodies in class, age, and geospatial contexts (ibid.: 188). He used the collection to demonstrate that the range of variation among Black bodies was not meaningfully different from the variation described in white bodies. He also pioneered the use of social data to show that social and historical context influenced how bodies developed and aged – poverty, he showed, inscribes Black and white bodies alike. Through the publication of this research, Cobb influenced a new generation of anthropologists and biologists, who also began to reevaluate the racist assumptions that had literally shaped what questions people asked, and what they saw in their data and evidence (Epps, Johnson, and Vaughan, 1993).

Cobb's project to combat the racial dimensions of scientific inequality formation in physical anthropology shows that, although receiving more attention of late from critical race theorists, we should not view Du Bois's case in isolation. The efforts of both men were simultaneously scientific and political in seeking to transform their home disciplines and broader society. Together, their cases illustrate how scientific inequality formation operates as a general process, one that may take particular shape within specific disciplines and mark certain institutions, such as Howard University, Atlanta University, and other historically Black colleges as potential sites of organized resistance. The remarkable achievements of Du Bois and Cobb, along with the efforts of contemporary scholars such as Morris and Watkins to recover the histories of their academic erasure and political organizing are themselves a form of insurgent network-building and organized academic resistance to the status quo. Together, the two cases offer a broader macro-historical critique of academic disciplines as they have organized knowledge and shaped social policy in the United States.

A history of knowledge told from the margins reveals disciplines to be powerful mechanisms that resist change (see 3rd Prescription for Practice) and thereby reproduce scientific and social inequality. As they matured into freestanding institutions during the first half of the twentieth century, sociology and anthropology reinforced a racial hierarchy of ideas and careers, cementing uniquely scientific forms of racial caste inequality into the basic structure of academic research and education. The contours of these historical absences remain with us today, shaping the contemporary politics of exclusion and ignorance in the social, biological, and health sciences (e.g. Brown 2007; Epstein 2007; Hatch 2016), as we illustrate in our conclusion.

Conclusion

Kettleman City is a small farmworker community of about 1,200 residents nestled in central California's San Joaquin Valley. It is a landscape that has been successively "cheapened" (Patel and Moore 2017), through centuries of racial and ethnic violence and nature and labor exploitation involving genocide and land appropriation as well as, more recently, industrial and agricultural development (see White 1991). In turn, the cheapening of the land and its people has spawned more than three generations of community environmental health and justice activism to combat the myriad chronic exposures to toxic hazards that people living in Kettleman City face on a daily basis (Cole and Foster 2001).

Chronic environmental exposures are so intense there that during an eighteen-month period between 2007 and 2009, eleven babies were born with severe birth defects; three of the infants later died (Leslie 2010). The bereaved families of these children, their neighbors, and two environmental NGOs suspected that pollutants from a nearby landfill were the primary cause and asked the State of California repeatedly for help with investigating the apparent birth defect cluster. According to environmental sociologist Lauren Richter (2017), despite a groundswell of community activism and media attention that involved a community health survey and extended over several years (Leslie 2010; Kumeh 2010), meaningful assistance from the state's public health department, environmental protection agency, and birth defects monitoring program never fully materialized. When it finally did render a verdict after years of delay, limited and belated data gathering, and reductionist analysis, "the state's investigation determined that the town's birth defects varied in type, concluded that the cases must be unrelated to one another and thus had no identifiable environmental cause (Richter 2017: 6–7).

As Richter describes it, the state's non-response is particularly troubling because Kettleman City is an "environmental sacrifice zone" (Lerner 2010), a place where "nearly every day of the year, people living and working here are exposed to air quality *known* to harm human health" (Richter 2017:6; emphasis in original). Water is so contaminated that the town's economically disadvantaged residents – 96 percent identified as Hispanic in the latest U.S. Census – must rely on bottled water for daily use (ibid.) And no wonder. The community is surrounded by industrial-scale agriculture and dairy production facilities and unconventional oil and gas extraction infrastructure and lies within a few miles of the U.S. West's largest Class I hazardous waste facility. These activities and related heavy diesel truck traffic put residents at risk for exposure to a range of toxic substances including pesticides, gas fumes, PCBs, arsenic, and asbestos (Kumeh 2017). State agencies and university scientists had previously studied these hazards and the risks they pose to human health. Yet even with so much already known about the environmental hazards swirling around this community, the birth defects cluster remains a mystery. Is there some vital piece of scientific knowledge that is missing? Or, as Richter's analysis suggests, does racial and economic inequality in the San Joaquin Valley combine with institutional features of regulatory and university science to make invisible what everyone who lives in Kettleman City – residents who themselves are largely invisible to government and university scientists – already "knows"? Or both?

We are ending this chapter as we began it, with an environmental puzzle. In Kettleman City, as in post-Harvey Houston, government regulatory inaction generated new types of scientific absence. In Houston, a national hub of oil and gas industry, environmental agencies decided not to collect new, high-quality data on airborne contaminants in the context of *acute* chemical releases caused by hurricane damage. In Kettleman City, in the heart of California's agriculture and oil production, environmental agencies similarly decided not to pursue studies of an infant disease cluster in the context of *chronic* chemical releases from ongoing drilling, refining, transportation, and farming. Involving different regions, different populations, and different types of (acute vs. chronic) exposure to toxic chemicals, the two cases share features with other examples presented in this chapter in support of our central thesis that systemic productions and articulations of absence are indelible features of scientific inequality formation.

Across different cases spanning continents and centuries, we see the power of absence to build and reinforce scientific inequality formation and the social mechanisms that efficiently and invisibly inscribe and institutionalize it. We have identified some of these mechanisms, although there are certainly others. They include cultures of decision making that structure whose knowledge matters and whose doesn't, cutting short opportunities to generate new knowledge falling outside prescribed routines and normative expectations. In the implicit assessments of risk that weigh the costs of public and environmental harm as less important than the economic costs of regulation and less important than the cultural and political costs of dismantling racial hierarchy, we see that absences in science do not just happen, but are created – sometimes intentionally, sometimes unintentionally – by institutionalized mechanisms of exclusion built in to the structure of academic, medical, and regulatory science. And we see how, with frightening efficiency, the exclusion of people erases, disallows, or discredits existing knowledge practices and creates barriers to scientific innovation, whether among eighteenth-century West Indian midwives and healers, twentieth-century African American academics, or NASA scientists offering to help understand the impacts of Hurricane Harvey in this century. We see how the exclusion of data, such as a research sample of Black people's skeletons, reinforces social inequality but also impedes the organization of resistance to scientific forms of racism. We see how patterns of epistemic dominance, guided by the twinned ideas of objectivity and universalism, excludes ideas that percolate in from the margins, be they Native American STEM students, the path-breaking studies coming from W.E.B. Du Bois' Atlanta Sociological Laboratory, or the chronically poisoned residents of Kettleman City.

As we have argued, absence constitutes an invisible scaffolding for scientific inequality formation. Asking absent-minded questions about science and inequality allows this invisible infrastructure to come more clearly into view, the better to see how absence is yoked to social power, including the power to exclude, erase, forget, neglect, ignore, deny, and suppress. At the same time, the power to hide knowledge can operate as a form of power and agency among the excluded, oppressed, and otherwise marginalized groups. In the next chapter, we build on the idea of knowledge/ignorance as a form of resistance in considering the prospects for collective change in scientific inequality formation in the context of deepening social inequality and ecological crisis.

4

Challenging
Scientific Inequality Formation

In June of 1974, a letter signed by eleven eminent biologists appeared simultaneously in three leading scientific journals[1] calling for a global moratorium on experiments using recombinant DNA, or rDNA. The letter's genesis lay in a contentious discussion that had unfolded at the Gordon Conference on Nucleic Acid Chemistry convened in New Hampton, New Hampshire, the previous year. There, Stanford molecular biologist Paul Berg announced that his laboratory had spliced together DNA strands from *E. coli* bacteria and Simian Virus 40, a virus known to cause tumors in rats. The spliced DNA not only constituted a new life-form, but also was capable of replication. Berg's announcement was cause for excitement, but also concern. "Although such experiments are likely to facilitate the solution of important theoretical and practical biological problems," the letter noted, similar gene-splicing experiments "would also result in the creation of novel types of infectious DNA elements whose biological properties cannot be completely predicted in advance" (Berg et al. 1974). Wary that the hybrid rDNA molecules constituted an as-yet-unknown risk to public health and ecosystems if they were to escape from laboratories, the authors urged the broader scientific community to refrain from making or using them "until attempts have been made to evaluate the hazards and some resolution of the outstanding questions has been achieved" (ibid.). A few months later, in February 1975, dozens of scientists and a smattering of lawyers, journalists, and government officials gathered at the Asilomar Conference Center in Pacific Grove, California, to do just that.

The four-day International Congress on Recombinant DNA Molecules or more commonly, "Asilomar," provided a forum for discussion and debate among the roughly 150 people in attendance on the question of whether and under what conditions the worldwide moratorium should be lifted and experimental rDNA research resumed. Uncertainty was high. Would people infected by the genetically engineered DNA get cancer? Would the new bacteria turn innocuous microbes into human or animal pathogens? Would they produce harmful toxins? Would they be resistant to antibiotics? Peering across a newly opened knowledge frontier, no one had answers or clear insight.

Self-interest was also in abundance at Asilomar. As Berg later recounted, "I was struck by how often scientists willingly acknowledged the risks in other's experiments but not in their own" (Berg 2008: 290). Some worried that such self-serving arguments would undermine public perceptions of the legitimacy of the group's decisions and that "government interference or even legislation would follow" (ibid.). Yet, throughout the often-heated discussions, conveyed to the news-reading public by the journalists in attendance, the scientists agreed to lift the moratorium and resume rDNA experimentation, but under a new set of guidelines designed to minimize laboratory risk.[2] Soon after, the risk guidelines developed at Asilomar became enshrined as official doctrine in the United States and elsewhere (Krimsky 1982), and they remain the basis for biosafety regulations governing genetic engineering research today.

With safeguards in place, rDNA research resumed, helping to pave the way for numerous technological and commercial developments in agriculture, biomedicine, and biotechnology. These innovations – which include gene therapies, animal and human cloning, genetically modified crops, vaccine development, and commercialization of genomic information – have benefited many, but they have also concentrated capital, increased ecological risks, and widened gaps between rich and poor, with controversy and social protest often following closely on the heels of these innovations (Barinaga 2000; see also Franklin 2007; Kinchy 2012; Parathasarathy 2017; Schurman and Munro 2013). In hindsight, Asilomar did not so much redistribute scientific power, as concentrate it, reinforcing rather than reducing scientific inequality formation. For readers who see mobilization of scientists as a powerful force for institutional and social change, Asilomar offers a cautionary tale.

This chapter builds from the idea that scientific inequality formation is a durable form of inequality, held in place by the historical intertwining of knowledge production with profit, as described in Chapter 2, and by the invisible structuring of absence, as described in Chapter 3. Its obduracy also derives in part from the deeply conservative nature of scientific institutions, where tradition, stability, and order are justified as wellsprings for novel ideas and innovation (Abbott 2001), and where values of objectivity and professional autonomy are mechanisms for insulating science from politics and cultural ferment. Paradoxically, the institutions and values that feed a continuous stream of innovations in knowledge and technology also make challenges to scientific inequality formation unlikely to result in rapid, substantive, and lasting change – as we will see in the cases presented below.

Lessons from Asilomar

The story of Asilomar is well known, although the lessons that life scientists, risk analysts, and STS scholars draw from the case can vary. Some, including Berg and other Asilomar veterans (Frederickson 1991; Singer 1977, 1979), celebrate Asilomar as a model of scientific self-governance. In this view, scientists responded to the inherent risks of new technologies early and openly, illustrating scientists' ability to solve problems of their own making. Others have focused on the internal processes that unfolded at Asilomar, noting how amid heated argument, voting emerged as an important, if uncommon, mechanism for science decision-making (Guston 2005). Still others have studied Asilomar as a case of technological controversy, an opportunity to examine the relationship between experts, the press, and the broader public (Nelkin 2001). As Sheldon Krimsky (1982) has argued, the scientists at Asilomar gathered under a cloud of growing public distrust in science, spurred in the previous decade by technical disputes related to the war in Vietnam, nuclear proliferation, and rising awareness of environmental problems. The larger political conflicts strained the public's social contract with science, shaping scientists' response to the dilemma of risky science.

For us, Asilomar also raises important lessons about the potential of scientists, acting together, to confront asymmetries in science and society and to redistribute scientific power in lasting, substantive ways. This is asking a lot, we know, yet this view is consistent with our ongoing line of argument. We have argued somewhat cynically in earlier chapters that the power imbalances constituting scientific inequality formation are stubbornly obdurate because the cultural and social projects that structure them tend to be inward looking and conservative, tending toward order and away from uncertainty. For example, systems of peer review for journal publication can operate as a gift economy that rewards and sanctions adherence to traditional norms (Hagstrom 1965; Kaltenbrunner, Birch, and Amuchastegui 2022). Disciplines, anchored in university departments, reproduce epistemic cultures (Kuhn 1962) as well as Ph.D. labor pools (Kohler 1990). Steady streams of financial and other resources, including ideological support from state and market actors, hold these structures in place. And disciplinary divisions in science additionally ensure that change, if and when it does come, diffuses more easily within disciplines, than between them. Undergirding all of this are some very deeply entrenched values that guide and give meaning to scientific practice. One such value is objectivity, described in Chapter 3, which privileges scientific knowledge above all other ways of knowing. Asilomar highlights another value that is also integral to reproducing scientific inequality formation: professional autonomy. This is the belief that in order to produce objective knowledge, the scientific community must govern itself, relatively free from outside influence (Merton 1973). On balance, scientists' embrace of professional autonomy reinforces scientists' own professional interests, reproducing scientific inequality formation.

The vigorous defense of scientific autonomy was on full display at Asilomar. A small group of elite biologists put forward a plan to regulate themselves and allowed the return to scientific business as usual. Their strategy, if that is what it was, worked. It also proved useful for avoiding outside regulatory pressures from government agencies and for reasserting the social authority of science in the context of declining public trust. Moreover, the Asilomar researchers did so in a burst of "scientist activism" (Allen 2003), an instance of academic researchers putting aside daily research tasks to take action collectively in furtherance of shared political goals. Specifically, in this case, the collective defense of scientific autonomy and preservation of science's cultural authority in the public square (Gieryn 1998; Habermas 1973). In this respect at least, Asilomar undergirds our cynicism. It is an example of scientists mobilizing to reproduce, rather than redistribute, scientific power. By expanding the capacity of scientists to regulate themselves, Asilomar activists reasserted their social authority to make judgments on socially divisive issues in the absence of broader participation from policy makers and members of the public.

Reflecting on those efforts some three decades later, Paul Berg (2008: 291) expressed doubt that conferences like Asilomar could achieve similar results today.

> In the 1970s, most of the scientists engaged in recombinant DNA research were working in public institutions and were therefore able to get together and voice opinions without having to look over their shoulders. This is no longer the case – as many scientists now work for private companies where commercial considerations are important ... A conference that sets out to find consensus [on socially contentious issues] ... would, I believe, be doomed to acrimony and policy stagnation.

Berg's pessimism about the ability of scientists to collectively regulate scientific research may be warranted. Neoliberal ideology and policies were just getting started in the 1970s and federal legislation that would prove critical for neoliberalizing science in the United States was still some years off.[3] There is little doubt that Asilomar scientists' implementation and swift retraction of the moratorium on rDNA research helped power the broader commercialization of science, as we described in Chapter 2. But it is also likely to have intensified conditions for technological conflict and public challenges to scientific authority going forward – dynamics of the neoliberal era that David Hess (2016) has described in terms of "epistemological modernization." A contemporary comparison, modeled after Asilomar, is suggested by current efforts to "pause AI" (e.g. Future of Life Institute 2023) – a goal that, if achieved, could destabilize scientific inequality formation. The movement has generated media attention and public debate, but thus far no binding agreement to halt large-scale artificial intelligence experiments has emerged.

In the meantime, Berg's thesis remains an open question. What opportunities does the current political landscape in science hold for challenging – and substantively altering – scientific inequality formation in ways that reduce rather than reinforce inequality? What conditions would such a transformation require? What social and scientific forces would need to be marshalled to bring real and lasting improvements in the science–inequality equation?

Is Scientist Activism the Answer?

As we noted in our Introduction, scientists are people too (see 4th Prescription for Practice), with multi-dimensional lives that include political values, biases, interests, and commitments. So, if the term "scientist activism" strikes some as an oxymoron, it is only because these all-too human traits are routinely denied to people calling themselves scientists.[4] Scientists inhabit occupational worlds organized by a nested maze of socially constructed yet deeply institutionalized and historically embedded categories, identities, and boundaries defined by the particular cultures of universities, departments, disciplines, theoretical traditions, and the like. In that context, crossing the line(s) between politics and science often has real-world consequences for these individuals, the knowledge they produce, and the institutions they serve. Scientist activism can boost some scientific careers, but it can end them as well. Despite the risks, history is replete with examples of "dissident" scientists who have transgressed taken-for-granted boundaries and pushed against established norms, using politics to intentionally disrupt scientific business as usual (Delborne 2008): J.D. Bernal, Rachel Carson, Noam Chomsky, Barry Commoner, Angela Davis, W.E.B. Du Bois, Albert Einstein, and Robert Oppenheimer are some notable examples among many, many others. Dozens of STS studies profiling the activities and experiences of individual researchers who engage with politics and social movements *as scientists* also bear this out.

Take Wilma Subra, a former food chemist and environmental scientist, whose work with communities in Louisiana's infamous "cancer alley" is described in research by Barbara Allen (2003; 2004). Subra's efforts on behalf of communities poisoned by industrial wastes in Louisiana and across the country has earned her renown in environmental justice circles, opprobrium in the petrochemical industry, and criticism from many in the scientific community (Marsa 2015).

Through her own environmental consulting firm, and as technical director for the Louisiana Environmental Action Network, Subra has spent decades collaborating with local residents to conduct air and water testing, obtain chemical analyses, interpret data from government reports, advocate in the media, and challenge industry and government agencies, pressing for changes in the state's lax environmental regulation and enforcement system (Allen 2003). As much as Subra's communication, organizing, and technical skills have been lauded through prestigious awards and board memberships, her activism on behalf of the communities has also provoked anger and violence from others, including several attempted burglaries of her office and at least one drive-by shooting (Mullen 2020).

As an independent business owner and entrepreneur, Subra has operated as an outsider in research and policy settings. While her position in the private sector has not afforded her the legal and police protections that university and government employees might receive when faced with similar types of violence, she has also enjoyed a level of independence in pursuing a career centered on environmental health and justice activism that most chemists working in academic and government jobs do not. Using this independence – and driven by a strong moral commitment to the communities she serves – Subra has forged a distinctively successful practice that blends scientific research, confrontational protest, and media relations to force environmental action from corporate polluters and regulators, in part by "actively constructing networks and alliances that intentionally blur the boundaries . . . between science and politics" (Allen 2004: 437).

Another case highlighting the clear benefits and risks of scientist activism, even among eminent scientists occupying politically powerful academic and government positions, is Dr. Andrés Carrasco (1946–2014), of Argentina. A high-ranking molecular biologist at the prestigious National University of Buenos Aires, in the early 2000s, Carrasco directed the university's Molecular Embryology Lab and led the Research Department for the National Ministry of Defense. In response to anti-pesticide activists who were beginning to voice concern about the negative health impacts of human exposure to glyphosate pesticides from intensive spraying on commercial soybean farms (Arancibia 2013), Carrasco carried out an experiment to test the activists' claims. His study validated their concerns, finding that glyphosate exposure caused cranial, cerebral, intestinal, and cardiovascular malformations in frog embryos and likely caused developmental abnormalities *in utero* in humans as well. His study was the first of its kind in Argentina. But what transformed Carrasco into a scientist activist, virtually overnight, was his decision to publish the study results, not in a scientific journal subsequent to the standard peer review and revision, but in the country's largest daily newspaper, *Pagina 12* – written in Spanish, not English – a clear transgression of scientific norms. Millions of Argentinians read the frontpage story, sparking a national conversation and marking a critical inflection point in the country's rapidly growing anti-pesticide movement (Frickel and Arancibia 2022).

Carrasco's research[5] not only supported the movement's claims, but directly challenged the legitimacy of the government's dominant economic development policy, an industrial agribusiness model which involves production of genetically modified soy requiring intensive use of glyphosate-based pesticides and herbicides (Leguizamón 2014). Despite his professional reputation and position, Carrasco became the target of fierce "backfire" from officials in national government, mayors, academic administrators, industry, and the U.S. embassy (Martin 2007). His work was censored and he was harassed, but he was also threatened with physical and property violence, robbed, and ultimately forced to resign his government and academic positions (Arancibia and Motta 2019). His illustrious scientific career sabotaged, Carrasco devoted the remaining years of his life to radical protest and activism, a popularly celebrated scientist who "chose to walk alongside peasants, fumigated mothers, and peoples in struggle" (Aranda 2024).

Is the sort of scientist activism illustrated by Subra and Carrasco the answer to overcoming scientific inequality formation? Not likely. Or at least, not in isolation. Subra's work has been a consistent force for change in southern Louisiana for decades, but the state's dependency relationship with the chemical and fossil fuel industries in this "polluter's paradise" remains as formidable as ever (Baurick, Younes, and Meiners, 2019; see also Lerner 2005; Roberts and Toffelon-Weiss 2001). Regulatory officials' power to decide when, where, how, *and whether* state and federal agencies enforce existing rules (never mind creating new, more stringent ones) is undiminished (Ottinger 2013a). Subra's dissident science produces occasional smaller victories, but the larger war against industrial pollution and environmental injustice has not fundamentally changed. "There's just so much to do," says Subra. "If we're not there as a pushback, then they're going to just run over us and destroy the environment and all human health in the name of economic development" (quoted in Marsa 2015). In Argentina, what began as dissident science in Carrasco's lab has blossomed into a vibrant interdisciplinary field of pesticide research (Arancibia, Frickel, and Annud Hannod, 2024) and a social movement that has mobilized hundreds of scientists and other professionals, including lawyers, journalists, pediatricians, pharmacists, and school teachers (Frickel and Arancibia 2022). Yet the movement's central demands for federal and state laws regulating pesticide use to protect human health and the environment remains unrealized, and spraying in the country's soy-producing regions and the consequent poisoning of rural and peri-urban settlements continues unabated (Berros 2024).

Without question, dissident scientists like Subra and Carrasco are doing crucial political and scientific work when they "stick their necks out" (Schnaiberg 1977) for social change. Necessary as that work may be, however, it is insufficient for fundamentally challenging scientific inequality formation. Subverting the dominant projects of scientific inequality formation will require a larger and more sustained *collective* response from scientists and their allies, raising questions about *how scientist activism is organized* and what strategies might most effectively build capacity across the science-politics divide. One way is through social movements organized by scientists, inside science.

Scientific/Intellectual Movements

Who better to change science than scientists? This idea has been a motivating logic for scientists throughout history and, as Neil Gross and Scott Frickel have argued, scientific/intellectual movements, or SIMs, have long been important mechanisms for reorganizing the epistemic terrain of academic research (Frickel and Gross 2005). Conceptualized broadly as "collective efforts to pursue research programs or projects for thought in the face of resistance from others in the scientific or intellectual community," SIMs are akin to social movements that emerge within science (206). They involve coordinated collective action in pursuit of some set of intellectual and organizational goals, such as enshrining a new theory or methodology, establishing a new research specialty or discipline, or carving out different intellectual spaces or professional turf. Claims for new ideas, new fields, and new boundaries mean displacing or replacing older ones. For this reason, SIMs will almost certainly inspire some form of opposition, and the contention that invariably unfolds among the competing camps makes SIMs inherently political insofar as they contest the existing distribution of scientific identities, resources, and power.

The Asilomar rDNA controversy possessed many of the trappings of a SIM, in fact. It was collective, coordinated, and contentious in its successful bid to extend genetic engineers' reach into science policy. But that case is probably better viewed as one episode within the longer arc of a more substantial SIM organized within the molecular life sciences to establish the modern field we know today as biotechnology (Bud 1994). A defensive tactic orchestrated by molecular researchers to preempt perceived threats to their movement's longer-term legitimacy, Asilomar effected changes that ultimately reinforced scientific inequality formation. Because our goal is to appraise the potential for SIMs to effect changes in science and society that *reduce* inequality, a better illustration may be the case of genetic toxicology.

Genetic toxicology is an interdisciplinary environmental health science that investigates genetic and chromosomal effects of exposure to "environmental mutagens" such as radiation and some viruses, but especially chemicals (Wassom 1989). Understanding the mutagenic potential of a wide range of naturally occurring and synthetic chemicals has broad implications for applied research and policy for human health, risk assessment, and pollution prevention (Preston and Hoffman 2001). Geneticists had been using chemical mutagens as research tools since the 1940s and by the mid-50s had developed a basic understanding of how mutagenic processes worked. Yet, linking mutagens to public health and environmental risk did not occur until the late-1960s when a small group of geneticists, toxicologists, and pharmacologists, most of whom worked in U.S. government agencies and national laboratories,[6] founded the Environmental Mutagen Society (EMS). Created "to encourage interest in and study of mutagens in the human environment, particularly as these may be of concern to public health" (News Release, March 1, 1969; quoted in Frickel 2004a), throughout the 1970s, EMS operated as an organizational platform for scientist activists in constructing chemical mutagens as environmental problems and creating societal demand for genetic toxicology knowledge.

Coinciding with the rising tide of environmental awareness and activism, the movement to establish genetic toxicology drew on environmentalist discourse to make its case. Echoing Rachel Carson's observation in *Silent Spring* that "genetic deterioration through man-made agents is the menace of our time, 'the last and greatest danger to our civilization'" (Carson 1962: 186), genetic toxicology activists framed chemical mutagens as "genetic hazards" that damaged germ cells, threatening the long-term health of human populations and altering the course of human evolution. Representative of this framing strategy, a public lecture by geneticist Frederick J. De Serres warned his audience:

> What supreme arrogance and folly it would be for us to imagine that we are immune to the dramatic and deadly consequences of environmental chemical pollution and to sit back and do nothing about such a serious problem! If we do nothing we may well find ourselves in the position of the California Brown Pelican – just another animal on the road to extinction! (De Serres, mimeographed lecture [no date]; quoted in Frickel 2004a: 98)

De Serres was not alone in this. Fueled by the urgency to prevent a "genetic emergency" (Crow 1968), genetic toxicology activists mobilized rapidly. By 1972, the EMS roster listed 452 dues-paying members, published the first three issues of the *EMS Newsletter* and organized the Society's first annual conference. Published research on chemical mutagens during this period spiked 500 percent, prompting EMS to create a committee to develop a registry of chemical mutagens which would facilitate researchers' and policymakers' access to the new knowledge. Using ad hoc committees to develop programing, frame research agendas, and engage with policymakers was an organizational strategy for mobilizing EMS members into the discipline-building project and expanding the nascent field's political reach. For example, "one committee, convened in 1972, was charged with 'extending [the Delany Clause] to mutagens and teratogens,' and another, created in 1975, looked into chemical protections for workers" (Frickel 2004a: 127).[7] Other committees conducted literature reviews targeting specific chemical classes, developed policy recommendations, or organized design workshops on mutagenicity testing methods at a time when university toxicology programs offered graduate students little if any genetics training (ibid.).

EMS leaders also took the movement global, promoting genetic toxicology in public and scientific lectures across North America, Europe, and Asia. In the process, they helped establish eight new EMS-related societies and four new journals. In the U.S., Congressional passage of the Toxic Substances Control Act (TSCA) in 1976 authorized pre-market screening for new chemicals. The Act established a formal market for genotoxicity data, with a provision for mutagenicity testing drafted by EMS officer and genetic toxicology activist Samuel Epstein. When the Environmental Protection Agency standardized mutagenicity testing protocols a few years later the regulatory rules were set, marking genetic toxicology's official arrival. Thus, within a decade,

> the movement transformed chemical mutagens from tools of research in experimental genetics into a ... global environmental problem. Although led by the seasoned geneticists who established the EMS, the campaign to create genetic toxicology was widespread and intensely interdisciplinary, reflecting the efforts of scientists working in academic, government, and industry settings whose training was rooted in more than thirty disciplines and departments ranging across the biological, agricultural, environmental, and health sciences. (Frickel 2004b: 272)

Genetic toxicology was a contentious project from the start. Intentionally interdisciplinary, the movement challenged disciplinary and methodological divisions in academic science. In the midst of Nixon's war on cancer, it sought to decenter carcinogenicity's starring role in human disease. It allied itself with a "new environmentalism" and extended the movement's preservationist logic to human DNA. An early "impact science" in Allen Schnaiberg's (1977) terminology, it implicated industrial science and engineering in creating a "genetic emergency" and, through mutagenicity testing, offered a way to surveil chemical production systems and products to protect workers' and consumers' health that in some ways presaged environmental groups' promotion of biomonitoring three decades later (Creager 2018). Amplified by a politics of prevention, it pressured government agencies to re-conceptualize chemical risk and regulation, a program that swiftly became enshrined in federal law under TSCA. Yes, "the scientist-activists who created genetic toxicology transformed the institutions that made and ordered environmental knowledge" (Frickel 2004a: 146). But in transforming environmental knowledge, did the movement thereby weaken the forces of scientific inequality formation? More is known about what substances are damaging human DNA and why because these researchers mobilized to reframe chemicals as genetic hazards. Are societies and nature less threatened by chemical trespass today as a result? The movement infused environmental values into experimental practice in some corners of genetics and in regulatory toxicology. But have those values, in turn, been institutionalized on a scale that overcomes any of the many challenges that movement activists confronted in the 1970s? Historical evidence of broader institutional impact is rather thin.

Consider, for example, that genetic toxicology activists once worried:

> . . . the cancer people are pulling the purse strings tighter when it comes to giving money for genetics. It is not right that mutation work should have to be a tail to the cancer kite. I think the time has come . . . to get a movement to support it started . . . (Joshua Lederberg to H.J. Muller 1950; quoted in Lederberg 1997)

Today, in many respects, mutagenicity is still cancer's kite tail. Carcinogenicity remains the dominant endpoint for regulatory toxicology and chemical risk assessment and in this context mutagenicity testing is used primarily as a reliable and cost-effective way to screen chemicals for potential carcinogenicity, among other diseases (Eastman et al. 2009). Generally, regulatory agencies consider mutagenicity a lower-priority outcome in national and international chemical safety programs unless is it explicitly linked to cancer. By the 1980s, efforts to identify mechanisms of mutagenesis began to compete with the study of "DNA repair" (Friedberg 2007), a developing line of inquiry somewhat at odds with the logic of prevention animating genetic toxicology activists a decade before. Research interest in studying the extent to which cells self-heal from chemical damage marked a stark contrast to activists' early concern about the potential for chemical exposure to produce heritable mutations that individuals would pass along to offspring, weakening the human gene pool over time. For some genetic toxicology activists, Nixon's focus on cancer rates, mortality statistics, and repair mechanisms had it exactly backwards. As EMS treasurer Marvin Legator put it, "While lethal and sub-lethal mutations will be rapidly eliminated from the population and need not be of great concern, non-lethal mutations will tend to persist through several generations, with the duration of their persistence being inversely proportional to the severity of their effects" (Legator 1970: 253). To counter such population-wide threats, one of the movement's more far-reaching aims, articulated by University of Wisconsin geneticist James F. Crow (1968), called for developing programs through global institutions to monitor the genetic health of human populations. Such programs never materialized, perhaps on fears of enabling a resurgent eugenics movement.[8] Today, instead, we have dozens of commercialized and monetized cousins, such as Ancestry.com and 23andMe that profit by selling the genetic information they collect to pharmaceutical companies for drug development (Nelson 2016), including mutation data. Mutations are big business, not the global environmental and evolutionary concern that once worried geneticists like Crow, Lederberg, Legator, and De Serres.

The regulatory reforms that the movement helped set in motion also have fallen short of initial expectations. TSCA has been roundly criticized as a policy failure (e.g. Vogel and Roberts 2011). Critically, the Act "grandfathered" in some 62,000 chemicals then in circulation and granted numerous exceptions for substances such as pesticides, tobacco products, nuclear materials, food additives, and drugs, which were already regulated under different existing laws (McLean 2020). These explicit limitations, in addition to numerous bureaucratic burdens and legal challenges, meant that prying chemical and toxicity data from industry under TSCA was slow and spotty at best. As late as 1997, an Environmental Defense Fund study of 3,000 high volume production chemicals in the TSCA database found it to be riddled by absences (see Chapter 3), with 70 percent of the chemical entries missing basic toxicological and genotoxic information (EDF 1997). A much-lauded 2016 amendment[9] requires test data for some existing chemicals and closes several loopholes in the original Act (Schmidt 2016). Yet data collection remains slow and, as we learned during the first Trump Administration, the new law is susceptible to executive branch rule-making powers that allow political appointees to overrule EPA scientist recommendations on regulatory decisions potentially putting "the public's health at greater risk than even under the old TSCA" (Denison 2018).[10] The state of mutagenic chemical threats under the European Union's REACH program[11] fairs little better. A recent assessment found that "the REACH information requirements will not provide sufficient information to conclude a substance is a Cat 1B mutagen and/or carcinogen" and that "requiring such information via a substance evaluation under REACH requires a large investment from the Member States and takes years" (Felter et al. 2021; see also Woutersen et al. 2018).[12]

The scientists' movement to create genetic toxicology was a reform movement that changed the way researchers and government agencies understood the connections between chemical pollution, human DNA, and public health. These were important changes, but the movement did not substantively alter the structures of scientific inequality formation in ways that would prevent or even stall a global "genetic emergency." As others have pointed out, SIMs confined to professional academic spheres are not likely to trigger far-reaching social or environmental changes (Arthur 2009; Waidzunas 2013). In part, this is because scientific institutions have developed to accommodate the very sorts of epistemic and organizational changes that SIMs demand. To survive and flourish, new research fields paradoxically must adapt to the institutional conditions that established fields have created and continually reproduce (Abbott 2001). So, while conflict and change may be endemic to science, SIMs are constrained by the very structures and values that condition their emergence. They do not threaten the status quo in fundamental ways.[13]

If projects to redistribute scientific power are unlikely to originate inside science as SIMs, perhaps we should look at science as it sometimes is practiced outside the halls of academic, government, and industrial science.

Community Science

In many ways, "community science" is a radical departure from the scientific status quo. The term describes "collaboratively led scientific investigation and exploration [that] addresses community defined questions, allowing for engagement in the entirety of the scientific process" (Dosemagen and Parker 2019: 24–25). Distinguished from traditional academic science, community science develops from the outside in, with scientific institutions ideally playing supporting roles on community-centered, and often community-led, research projects. Often distinguished from "citizen science," which traditionally has involved scientist-led research using crowd-sourced data collection efforts that can involve thousands of otherwise unconnected volunteers,[14] community science "emphasizes the community's ownership of research and access to resulting data, and orients toward community goals and working together in scalable networks to encourage collaborative learning and civic engagement" (ibid.: 25).

Case studies of environmental community science, which often depict scientists as politically and culturally complicated actors (see 4th Prescription for Practice), are now legion in STS. Indeed, we have already encountered a few examples earlier in this book. The Network of Free Seeds of Colombia's "seed schools" and the community health surveys conducted by medical students in soy-producing regions of Argentina that appear in Chapter 1 are projects of resistance grounded in community science models of organizing. So were the "insurgent intellectual networks" that W.E.B. Du Bois created in collaboration with students, teachers, ministers, and activists to collectively produce a sociology of race and racism reflecting the lived worlds and experiences of Black people in the United States, in Chapter 3. Also, Wilma Subra's environmental justice work in south Louisiana from the previous section of this chapter. Other studies speak to the broad political potential of community science strategies for identifying cancer clusters in Massachusetts (Brown and Mikkelsen 1997), air-quality monitoring in fenceline communities in California (Ottinger and Sarantschin 2017), community health studies in Italy (Allen 2018), soil contamination in Chile (Ureta 2017), and plastics pollution in the North Atlantic (Liboiron 2021). As the large body of case study research demonstrates, community science writ large is a democratization project aimed at claiming different publics' capacities and rights to produce credible, actionable knowledge that addresses actual harms and proposes workable solutions. Community science expands social norms about what it means to do science. Beyond knowledge production, its collaborative practices nurture networks of trust that connect institutional science professionals to the lives and needs of local communities, helping channel science toward more socially responsible and just ends.

Against the odds, community science seems to be gaining institutional traction. In the United States, government agencies such as the Environmental Protection Agency and the National Institute of Environmental Health Sciences are embracing principles of community science, expanding the notion of community "engagement" or "outreach" well beyond the agencies' historical focus on risk communication and science literacy. Shannon Dosemagen and Alison Parker (2019) report that newly afforded institutional legitimacy of community science supports a wide spectrum of change opportunities ranging from raising community awareness about environmental and health issues facing residents, to building baseline datasets that allow communities to track environmental and health changes over time, to informing regulatory decision making, standard setting, and enforcement. Another potential advantage of community science is the motivation it can create for structural and cultural changes "upstream" into scientific cultures and communities (Frickel et al. 2022). For example, our own research has shown how connections between MIT urban studies faculty and community organizers in Greater Boston in the 1960s and 1970s continue to influence the politics of community development and environmental activism in the region today (Porcelli, Frickel, and Niznik 2022). Other studies emphasize the emotional connections and caring practices that can emerge when epistemic attention is directed to "neglected things" (Puig de la Bellacasa 2011; Murphy 2015) and, we should add, to neglected people as well.

Yet, as institutional support for community science has grown, critiques have also begun to emerge. The most forceful set of criticisms to date come from sociologists Aya Kimura and Abby Kinchy, whose book *Science by the People* (2019) offers a clear-eyed assessment of the power differentials that often confront community science practitioners in the neoliberal era. In their view, "participation in scientific knowledge making [does] not necessarily enable people to challenge powerful institutions ... or achieve desired outcomes" and they ask "what hinder[s] these projects from having more transformational effects?" (3). In response to their own question, Kimura and Kinchy describe four different, but often interconnected, limitations facing community science as a challenge to scientific inequality formation.[15]

One limitation is these projects' foundational reliance on volunteers. Unpaid scientific labor often substitutes for the paid work that private sector and government researchers arguably should be doing (and in some cases were doing before neoliberal policies shrunk state workforces and pursued a strategy of deregulation). Volunteer-based science can also send a social message that "we can fix it" when, more realistically, legal and regulatory changes are necessary. Moreover, because different social groups have different abilities to expend time and energy doing volunteer work, the model may unintentionally exclude those unable to participate, introducing demographic inequalities into community science projects. A second limitation is the politicization of knowledge that is inherent (but not unique) to community science. Often seen as "tainted" by its own political commitments, community science projects face uphill battles in earning credibility from allies as well as opponents, as many published studies detail (e.g. Ottinger 2010). Relatedly, because data quality and assurance often are at the heart of the credibility contests surrounding community science, technical disputes about evidence can overshadow other social, economic, cultural, and political issues that contextualize community science approaches to problem solving. Paradoxically, while community science is arguably best suited for illuminating these broader power dynamics, their effects often hinge on narrowly framed technical questions of reliability, validity, and interpretation.

A fourth limitation that Kimura and Kinchy describe involves the localized nature of nearly all environmentally focused community science. This is a double-edged sword. On one hand, these projects are able to generate data specific to particular geographic and cultural contexts to inform place-based understandings and solutions. In this sense, the localized character of community science is exceptionally important because it responds directly to the needs of specific communities in ways that academic and government science often does not. For example, *Breathe Providence* is a community-based network of twenty-five continuously operating air monitoring stations in Providence, Rhode Island.[16] In building the network, organizers were guided by local social geographies to ensure that the monitors oversampled underserved, minority, and lower-income neighborhoods. The project fills an important regulatory gap by generating street-level air quality data that shows residents, policy makers, and regulatory scientists which pollutants are circulating where and when. In a city in which childhood asthma rates are well above the national average, residents' pressing health concerns cannot be fully understood using the state's air monitoring system, which was not designed to characterize local pollution variability and does not reflect community members' priorities.

On the other hand, Kimura and Kinchy argue, the local character of most community science can also mask structural inequalities that reproduce environmental and social harms across geographies and over time. This is especially true when community science is designed to address local instantiations of systemic problems or inequities. Community science is easily overlooked or dismissed as anecdotal by academic and government scientists, who are driven by generalizable knowledge that advances theory or satisfies policy. When this happens, as it often does – recall the state of California's non-response to the evidence-based health concerns of Kettleman City residents, in Chapter 3 – it becomes incumbent on community science projects to "scale up" to meet the problem where it is so often rooted: in the durable structural patterns of scientific inequality formation. Scaling up requires resources that local communities often do not have. It also requires coordination across communities aided by alliances with social movements, the media, sympathetic politicians, and others. These are difficult, but often necessary political and organizational challenges that extend beyond the knowledge that community science produces. When coordination challenges are unmet, even local victories can adversely impact the larger struggle by reinforcing public perception that community-based problems are unique and solvable at the local level, leaving the larger structural causes unaddressed.

Kimura and Kinchy's four criticisms of community science are as empathic as they are revealing. They are born from the two sociologists' own research experiences and shared commitment to community science as a form of STS practice. Consider

Kimura's (2016) analysis of Citizen Science Radiation-Measuring Organizations (CRMOs) that emerged across Japan following the catastrophic rupture of the Daiichi Nuclear Power Plant in Fukushima in 2011.[17]

CRMOs are citizens groups who established stations where residents could bring food and beverages to be tested for radioactive cesium contamination. These groups provided inexpensive or free testing using low-cost radiation detectors purchased by CRMO organizers and operated by trained volunteers. Part of a larger study of the gendered politics of food contamination following the Fukushima disaster, Kimura documented at least seventy-four CRMOs and interviewed representatives from sixty-five. The study offers a perspective on community science outside the U.S. and Europe and – importantly – presents community science in the context of a national ecological, public health, and political crisis. If our goal is to assess the potential for community science to challenge scientific inequality formation beyond the local level, this is a good one to consider.

Distrusting the government's claims about the nature and extent of contamination risk in the wake of the disaster, tens of thousands of Japanese citizens mobilized in the streets. For many, the protests signaled a growing legitimacy crisis for the so-called "nuclear village" of government officials, technocrats, and banking and business leaders controlling Japan's powerful nuclear industry. For their part, nuclear scientists offered repeated public assurances that the radiation release was under control, that radiation science could accurately assess the risks posed to the population and environment, and that food and water was generally "safe" to consume. Such pronouncements fueled public distrust in nuclear experts, whom the public increasingly perceived as "handmaidens of the government . . . unable to speak out against the government's positions" (Kimura 2016: 44). In this context, "many observers heralded CRMOs as part of a new wave of activism in Japan, an instance of progressive [community] science and rejuvenated social action . . ." (Kimura 2016: 105).

Yet, interviews and other research revealed something quite different. Rather than adopting science as a form of radical politics used to make explicit demands on the state for social change, Kimura found that most CRMOs employed science and testing as a *substitute* for political action that sought redress from the nuclear village. In the context of widespread public distrust of government nuclear policy and the nuclear village's response to the Fukushima disaster, what explains these groups' failure to challenge the scientific status quo?

Some of the answer lies in Japan's conservative political culture, shaped by the country's checkered history with progressive politics from earlier post-war New Left and anti-nuclear movements. More than in other liberal democracies, social movement politics is stigmatized in contemporary Japanese culture. With conformist pressure spiking after the nuclear disaster, activism cut sharply against a national political culture that "emphasized charity, volunteerism, and unity" (Kimura 2016: 120).

Another element of Kimura's explanation lies in the characteristics of the CRMO organizers themselves, few of whom came from activist backgrounds. Rather, many CRMO organizers identified as parents (mostly mothers) whose primary concern involved the health of their and others' children. Others, motivated by religious or communitarian beliefs, translated a similar ethics of care into providing support for refugees displaced from the disaster zone. Still others came from farming backgrounds or were otherwise involved in the food industry (for example, as restaurateurs or market sellers) for whom testing was motivated by economic and livelihood interests. And still others came to CRMOs as techie do-it-yourselfers interested in operating radiation detectors and tinkering with the technology to improve the machines' precision and accuracy. Although CRMOs practiced "small p" politics by rendering radiation visible to local residents who could collectively share the knowledge and compare community science measurements against the government's non-transparent assurances of safety, most were not motivated by politics, *per se*.

But the dominant factor that Kimura's analysis relies on to explain the "non-activism" of Japan's CRMOs in the aftermath of the 2011 nuclear disaster is the performative role of the "citizen-subject" in the era of neoliberalism. Neoliberalism constructs discourses of citizenship in historically specific ways, in accordance with the productive demands of the state and markets. In the Japan case, the ideal neoliberal citizen-subject is cooperative, collaborative, responsive and responsible to oneself, and forward-looking. And largely apolitical. Indeed, many organizers took pains to distance their CRMOs from the social protests and from the "activist" label and most did not organize with one another to establish larger, more sustainable, and more politically powerful CRMO networks.

According to Kimura (2016: 122), CRMOs were crystalized expressions of Japanese neoliberal governance and "science was an ideal tool to perform such helpful and useful citizenship." Illustrating a practical-minded complacency toward politics, one informant stressed that, "we devote ourselves totally to testing" (ibid.). Other interviewees described the core mission in their CRMO as serving the public through testing that was "good," "solid," "attentive," and "proper." Many shared the view that testing was a way to solve immediate problems – what foods are safe to eat? But testing also functioned to absolve the state of accountability and to rehabilitate the public reputation of radiation science as a source of politically neutral knowledge. In this way, CRMOs did double duty as a means "to rescue science from politics and also . . . to rescue citizens from politics" (Kimura 2016: 123).

As with other cases we encounter in this chapter, CRMOs did a lot of important work as Japan recovered from the nuclear disaster. They provided testing in places that lacked access to government testing stations. They generated tens of thousands of tests from across the country, and shared the results with local residents. The measurement data offered a level of transparency that contrasted sharply with the government's practice of notifying communities only when radiation levels registered above official safety limits. CRMO testing occasionally exposed mistakes and identified regulatory gaps in government testing programs, which excluded, for example, garden produce, foraged plants, and wild game. We shouldn't under-appreciate these and other contributions.

Yet, if CRMO practices were subversive as an informal government watchdog, they were not overtly political and did not directly challenge radiation science or the nuclear village. "Under the strong influence of neoliberal pressure for personal responsibility, testing [became] refashioned into a means for individuals to take responsibility for managing their exposure levels rather than holding the government and the nuclear industry accountable for contamination" (Kimura 2016: 125). Held together by shoestring budgets, by 2013 many CRMOs had begun closing up, the consequence of dwindling memberships and declining public interest amid a steady drumbeat of government assurances that radiation risk was steadily diminishing.

Conclusion

Back in Providence, researchers and students overseeing the *Breathe Providence* air monitoring network have been hard at work tracking in real time the major pollutants fouling the city's air. The monitors measure ambient concentrations of five priority pollutants at near-ground level: carbon monoxide (CO), carbon dioxide (CO_2), ozone (O_3), nitrogen oxides (NO_x), and fine particulate matter ($PM_{2.5}$), as well as tracking meteorological conditions including wind, temperature, and precipitation. Data are collected every minute, creating a near-continuous database of hyper-local air quality – air that city residents actually breathe. After correcting for measurement errors, these data are published on the organization's website allowing the public to study – and question, first-hand – air quality trends where they live, work, and play.

Since establishing the network in 2022, the community-focused project has already begun attracting attention from state agencies, the state's attorney general, community groups, and local media. Mindful of the tendency for grassroots community science to come under intense public scrutiny and criticism (Ottinger 2010; Kimura 2016), project organizers and students have worked to ensure that the network geography oversamples socioeconomically disadvantaged neighborhoods and that the monitoring equipment across the network is properly calibrated and field-tested (Fay 2023). Validating the data, in turn, allows for methodologically rigorous analysis of inter-urban trends and thus far has helped researchers identify hyper-local sources of air pollution and explain variability in pollution concentrations from neighborhood to neighborhood (Gendreau et al. 2023; Gendreau 2024). Other reports have documented the local impacts of Canadian wildfires (Farber et al. 2023) and a large fire that broke out unexpectedly (and illegally) at a local scrap metal recycling facility. The latter effort, prompted by a call from a community organizer, tracked the resulting plume as it moved through the city, sending $PM_{2.5}$ concentrations skyrocketing in its wake and prompting a district court judge to temporarily close the facility

(Hastings 2024; Lavin 2024). These and other "responsive analyses" are tailored to address specific concerns voiced by community members. The learning that takes place is then used by both groups to further develop educational programming for local schools and community groups and to develop policy recommendations for city and state officials. As of this writing, for example, city and state officials are considering *Breathe Providence* proposals to redesign road infrastructure to reduce diesel truck traffic through working class neighborhoods and to connect a citizen odor-reporting app into the Mayor's 3-1-1 city services hotline. In these ways, the project is lifting up community knowledge as a valuable source of environmental data and creating opportunities for socially robust local knowledge to generate higher-level changes in city and state officials' understandings and responses to air pollution.

Will *Breathe Providence* change air pollution science? One could argue that it already has. The project offers an implicit, but direct rebuttal of the state's air monitoring system, which is primarily designed to monitor regional-scale air pollution dynamics and to identify violations of air quality standards under the U.S. Clean Air Act. The *Breathe Providence* network is producing a different kind of knowledge, one that is legible to meteorologists and geochemists, but that follows a socio-logic of environmental equity and justice (i.e. one that is protective of underserved people where they "live, work, and play") rather than the legal logic of the Clean Air Act, which assumes equality of exposure risk across a national population. Imagine: a distributed grid of hyper-local air quality networks, one in every U.S. city, producing real-time data available to every resident and that aligns scientific data with their embodied experience of pollution.

This, perhaps, is where we can find some hope in challenging scientific inequality formation: in locally inspired efforts that do not simply replicate regulatory science, as the Japanese CRMOs aimed for, but augment, recalibrate, reorganize, and redeploy science to work for the people, in direct response to community needs. When this happens, the implicit critique of Science becomes explicit in practice, with the practice offering a way forward to a different kind and caliber of science; a science that is scalable through distributed scientist-activist networks, overcoming the insufficiencies of local knowledge and the intransigence of exclusive scientific norms; a science that is coordinated across

geographically disconnected places and institutional arrangements, overcoming the problems of scarce resources and organizational capacity. In short, a science that redistributes scientific
power by decentering Science.

Breathe Providence is not alone in marking this potential.
Other projects emerging through the work of Indigenous STS
and feminist STS scholars are building new kinds of environmental sciences that rehabilitate traditional relationships to land and
plants and bodies (see, e.g., Liboiron 2021; Murphy 2017b). In
Latin America, scientists, rural communities, and environmental activists are building new agroecological sciences that push
directly against the positivist, reductionistic model of agricultural
sciences inherited from the Green Revolution and institutionalized today in extractivist development policies, some of which
we encountered in Chapter 1 (see, e.g., Altieri 1995). Hacker
clubs, organized through distributed networks across the world
seek to build an "open science" that undermines the proprietary
logic of scientific knowledge by handing the means of scientific
production – the microscopes, telescopes, sensors, software, data,
and laboratories – over to the people (Albagli, Maciel, and Abdo
2015; see also Arancio and Dosemagen 2022). A lot of different
projects are developing in a lot of different places, but we're not
there yet. What will it take to make a critical, substantive, and
scalar difference in the distribution of scientific power?

Provocation:
Toward a Deeply Adapted Science

In 2018, Jem Bendell, then a professor at the Institute for Leadership and Sustainability at the University of Cumbria (U.K.) self-published an essay titled "Deep Adaptation: A Map for Navigating Climate Tragedy." It went viral, quickly amassing more than one million downloads (perhaps you have read it?). The paper's controversial central premise was that it is "too late to avert a global climate catastrophe in the lifetimes of people alive today" and that, consequently, "we now face inevitable, near-term societal collapse" (Bendell 2018: 1, 2). Societal collapse indicates "an uneven ending of industrial consumer modes of sustenance, shelter, health, security, pleasure, identity and meaning" (Bendell and Read 2021: 2). In other words, a comprehensive breakdown of modern social systems that surpasses society's ability to rebuild to prior complexity and scale. In this context, "deep adaptation describes the inner and outer, personal and collective responses to either the anticipation or experience of societal collapse, worsened by the direct or indirect impacts of climate change" (ibid.). As Bendell clarifies in the original piece, deep adaptation is not an extension of climate adaptation research and policy, which assumes that contemporary social institutions can manage the coming risks. Deep adaptation begins from the assumption that they cannot.

The chilling argument and program of action has prompted swift and withering criticism from climate scientists and activists, who contest Bendell's interpretation of climate science and reject the resulting "doomism" that many see as detrimental to climate mobilization and action (e.g. Nicholas, Hall, and Schmidt 2020). Writing on deep adaptation is also, thus far, largely devoid of social science thinking. Nevertheless, the article has helped to launch a new field of study that proponents are calling "collapsology" and to fuel ongoing discussion, debate, and analysis (Bendell and Read 2021; Cassegård 2024; Saltmarsh 2022; Servigne and Stevens 2020). Thus far, however, the provocation has prompted little discussion of what kinds of science might survive into a new Dark Age.

We do not (and do not need to) accept Bendell's apocalyptic premise of inevitable societal collapse to recognize the urgent threats – as well as potential opportunities – that the climate crisis, economic fragility, and ethno-nationalist populism are posing around the world today. Some see these unsettled times as signaling the end of neoliberalism and the emergence of a new (although likely more authoritarian or "post-liberal") political order (Gerstle 2022). Others forecast political disintegration (Turchin 2023). Whatever the ultimate outcomes, the forces of change that these compounding ecological, economic and political crises represent will almost certainly shake what we know about the world and how we know it. STS scholars are well-positioned to observe the impacts of compounding crisis on science and to consider what a more "deeply adapted" science might or should look like.

Even if wholesale societal collapse is not in humanity's near future (and we will know soon enough), erosion certainly is. Erosion is a process involving the "gradual reduction or destruction of something."[1] We are witnessing ecological erosion as glaciers melt, coastlines shrink, islands disappear, ocean currents slow, weather patterns reorganize, and extinctions mount. Analogously, for those who care to look, we are also witnessing erosion in many of the core social institutions that characterize most modern liberal democracies. These include the rule of law, free and fair elections, state protection from arbitrary or extra-judicial violence, and separation of church and state, all of which are under increasing pressure in many countries. Institutional erosion can also impact institutions of higher education and science. Expectations of state support for research and teaching, equal access to education and the products of research, and norms of tenure and promotion and academic freedom are weakening in many countries, compounding conditions of scientific inequality across and within nation-states.

In our experience, many climate scientists – including many of our social science colleagues who study climate change – are keenly aware of processes that we are labeling ecological erosion. Yet too often, it seems, researchers pay far less attention to kindred processes occurring in our own universities, colleges, laboratories, and disciplines, and remain naive to the ways that ecological and social-institutional erosion might couple and mutually reinforce one another (but see Rebecca Elliott's excellent 2018 essay on a sociology of loss). As diverse studies are showing, however, the climate crisis is placing ever greater stress on systems of disaster response and recovery (Pais and Elliott 2008; Shi et al. 2022; Griego et al. 2020), on insurance and re-insurance systems (Kousky 2019; Elliott 2021), on health systems (Parks et al. 2021; Young and Hsiang 2024; Kadandale 2020), on community and family structures (Wing et al. 2022; Helm et al. 2024).

More broadly, a world of intensifying storms, wildfires, floods, and drought deepens economic inequality (Howell and Elliott 2019; Pardy et al. 2024; Sheng et al. 2023), spawns military conflict (De Châtel 2014) and newly vicious forms of profiteering (Klein 2007)), and enables autocratic governments to gain political power (Rahman et al. 2022). Growing uncertainties associated with a rapidly changing climate system also threaten to undermine the predictive capacity of science grounded in comparisons of baseline data, historical patterns, and controlled experimentation (Beck 1995).[2]

This mounting challenge extends to the social sciences and to STS as well. Even efforts to mitigate the worst climate impacts by transitioning to alternative energy technologies and infrastructure are creating new environmental risks and societal dislocations that threaten to reproduce historical social inequalities even as they create new axes of inequality (e.g. Nsude et al. 2024; Ottinger 2013b; van Bommel and Höffken 2023). Given all of this, it seems to us a mistake to assume that the institutions of science and learning that we know today (warts and all) will remain robust and generative in the face of mounting and plural forces of ecological and social erosion. Why would they?

Indeed, science is unlikely to continue operating as it is accustomed to doing under material and cultural conditions marked by steepening institutional instability, resource scarcity, and political unrest. As processes of ecological and social erosion gain momentum, many scientists and their organizations will come under increasing pressure to produce more and better knowledge, even as the material, institutional, symbolic, and human resources that science requires become less dependably available. The social authority that scientific expertise still enjoys in most segments of society is not assured, either, as signaled recently by the U.S. Supreme Court's decision to overturn a forty-year precedent (the so-called "Chevron Doctrine") which gave federal agencies authority in interpreting ambiguous laws when those laws were disputed in court. Now, decisions about contested air-quality standards or vaccine development protocols will be made by judges rather than expert committees of the Environmental Protection Agency or the National Institutes of Health (Furlow 2024).

Will scientists, facing erosion of their institutions, authority, and professional livelihoods mobilize as they have many times before to defend the status quo, reproducing existing inequalities that structure contemporary science in the process? Or, will they take a different path, seizing opportunities that erosion creates to build new ways of doing, supporting, and circulating science that decouples research and knowledge from the structures of disadvantage and privilege that have fed and nurtured it for decades, if not centuries? Either way, erosion is likely to prove instrumental in weakening the social and epistemic authority of science. Will it also loosen the structures of privilege and disadvantage that currently undergird scientific inequality formation? Will it create social conditions for the emergence of "deeply adapted" sciences that embrace value-commitments to justice and community and an augmented objectivity that acknowledges both lived experience and the heterogeneity of the sciences (Harding 1991; Latour 2018)? Such potentialities bear directly on the arguments we have begun to sketch in this book.

A deeply adapted STS would require, first and foremost, frank diagnosis of the problem, an honest reckoning of what is wrong with science. Our diagnosis from Chapter 1 is that "inequality is built-in to science." Political sociological analysis of this claim, grounded in attention to scientific inequality formation, offers a critical appraisal of structural conditions shaping scientific institutions, knowledge, and research. We provided a set of historically and geographically diverse cases to illustrate core features of the argument – framed in terms of profit, absence, and challenge. Together, the framework and the cases begin to explain in broad strokes why the science–inequality relationship is ubiquitous yet often invisible, dependably fluid, yet stubbornly durable.

How well our diagnosis stands up to scrutiny remains to be seen. It is incomplete and limited, to be sure. For example, there are certainly more dimensions to scientific inequality formation than those we chose to study. While we think those covered in Chapters 2–4 are fundamental, there is more to scientific inequality formation than political economy, structured absence, and organized resistance. Moreover, our attention to various axes of inequality focused primarily on the classic sociological trinity of racism, classism, and sexism, largely ignoring their intersectionality, never mind physical ability, neurodivergence, Indigeneity, sexuality and sexual orientation, religious belief, citizenship status, and national origin – all of which warrant inclusion in future studies of scientific inequality formation. Our selection of cases, therefore, is neither representative nor special. Scores of other cases would no doubt prove equally illustrative. Conversely, scores of different cases could weaken the argument we are building.

We also considered cases that somehow manage to span half a millennium – from alchemy to Asilomar – in a perhaps fanciful effort to anchor our argument in what French historian Fernand Braudel (2001: xxv) imagines in his observation that "History may not repeat itself, but it is all part of a single fabric." Testing our argument's internal logic and its broader scope conditions are crucial for understanding whether our diagnosis withstands additional empirical research and where theoretical refinements, additions, or excisions are needed. The five "prescriptions for practice" enumerated in Chapter 1 and spotlighted here and there across the other chapters are small but we hope meaningfully pragmatic contributions toward this end, recognizing that the project of challenging scientific inequality formation begins with identifying and understanding it.

A deeply adapted STS will need theories and research methods that are up to the task. The social sciences developed in late nineteenth and early twentieth centuries in response to the productive forces of urbanization, industrialization, and bureaucratic organization of states and markets. Sociology, in particular, is built to study how societies modernize and develop, not erode and devolve. STS is far younger and for the most part attuned to studying the sciences as already established, well-oiled epistemic machines (i.e., laboratories, disciplines, technological systems, regulatory regimes). STS researchers will need different tools for studying how these machines break down and how to engineer different ones to better serve the needs of unevenly vulnerable human and more-than-human communities facing multiple, compounding threats.

On the other hand, simply calling for "more research and methods" in isolation is a non-starter. At all costs, a deeply adapted STS must avoid the trap of scientific "gaslighting" (Hatch 2022: 2), which justifies research as a substitute for social and political action. We think that one way to avoid the trap is to adhere to core STS principles of symmetricality and reflexivity (Bloor 1991). The ideas that all knowledge claims require social explanations (symmetricality), including claims produced by STS researchers (reflexivity), infuse this book. As we have tried to show, STS's asymmetrical inattention to inequality over the past half century creates a looming epistemic absence that has motivated our own work and adds collective urgency and the moral clarity of purposeful action to social studies of science. If STS has been complicit in reproducing scientific inequality, as we have suggested, then as individual practitioners and as members of larger research communities we are obliged by the principle of reflexivity to take inequality seriously as a core topic of research and teaching and to Do Something About It.

What that Something should be is obviously not for us to recommend, but specific courses of individual and collective action are numerous and available. They include the types of scientist activism, SIMs organization, and community science projects described in Chapter 4, but extend beyond these projects as well.

Sometimes, conducting research can be a meaningful form of social protest. In that spirit, consider questioning disciplinary authority and rethinking what it means to do "good" or "meaningful" or "just" science. Innovative models for this are Liboiron (2022) and Ureta (2021). Or, follow W.E.B. Du Bois and develop insurgent intellectual networks as a strategy for building scientific capacity in communities that lack it and find other ways to integrate science in civil society (Frickel 2015). Organize colleagues to shift tenure and promotion requirements to credit faculty for engaged scholarship. This will create openings for graduate students to legitimate dissertations designed to address community concerns and is a form of refusal to the "publish or perish" mentality of most universities and colleges. Consider publishing less to make time for work that is more consequential for vulnerable communities than for your discipline. Host a salon. Organize a community training session. Join a movement. Or start one (Moore 2008). All of these projects, and more, represent small but meaningful ways to create deeper social connections in and beyond science. Against admittedly long odds, efforts to coordinate and connect disparate projects provide some hope for forging resilient institutional alignments that confront scientific inequality formation by embedding science in civil society rather than walling science off from the societies that STS – and the sciences writ large – ostensibly serve.

Our brief exploration of science and inequality tempers whatever optimism we had hoped to bring to this conversation. A political sociology of science trained on empirical examples from the past suggests that science is unlikely to rise as a unified political force to alter the social and epistemic conditions feeding and protecting its most elite actors and institutions. Then again, as ecological and social erosion gain momentum, history will become less reliable as a guide to the future, giving us cautious hope that real change – and actual equality – remains possible. Having for so long "followed scientists around" (pace Latour), perhaps it is time for STS to take a lead role in generating actionable science that reconfigures knowledge production toward more equitable ends. At the very least, we should work to enhance and broaden ongoing efforts to understand and challenge the structures that hold us back.

Notes

Acknowledgments

1 Homer, *Odyssey*, Book 1, line 183; https://www.perseus.tufts.edu
/hopper/text?doc=Perseus%3Atext%3A1999.01.0136%3Abook
%3D1%3Acard%3D178.

1 Science, Society, and the Paradox of Inequality

1 Counter examples exist as well. Bioterrorism and the climate crisis
may make the world less safe, prosperous and united, but these are
understood to be failures of democracy, not failures of science *per se*
(Merton 1973).

2 An admittedly crude but nonetheless meaningful measure of this
claim, the terms "equality," "inequality," "justice," and "injustice"
do not appear in the subject indexes of any of the four published
editions of the *Handbook of Science and Technology Studies*, sponsored by the Society for Social Studies of Science. The term "environmental justice" is indexed in the most recent, 2017 edition.

3 Foundational work in this area includes Duster (1982); Harding ed.
(1991); Haraway (1989); Rose (1975); Shim (2005); Reardon (2004);
Hayden (2004); and TallBear (2013); as well as scholarship by scientists, including physical anthropologists, who since the 1930s were
using political arguments and scientific evidence to refute scientific
racism (Cobb 1936; Lewontin, Kamin, and Rose 1984; Selcer 2012).

4 Political sociological approaches to science and technology origi-
nated with British sociologist Stuart Blume (1974). Our own efforts
to rehabilitate and revitalize Blume's framework (Frickel and Moore

2006b), have since generated several additional statements with other colleagues that refine and extend the project. The statements include arguments for historicizing science in the context of neoliberal globalization (Moore et al. 2011), using field theory to study science as a product of cross-currents running between academic, state, industry, and civil society fields (Frickel and Hess 2014; Hess and Frickel 2014), and emphasizing the role of social movements and other "mobilized publics" in shaping the institutional and epistemic politics of science (Frickel et al. 2010; Hess et al. 2016). The book you are currently holding, on science's timeworn interconnections to social and inequality, is the latest. Taken together, these statements build toward an increasingly nuanced argument about what is "political" about science.

5 Continental geography itself is a European categorical convention.

6 Our view of the politics of science differs from that of political scientists and some sociologists, who tend to view inequality as a set of outcomes of politics-as-democratic-decision-making or politics-as-contention (see Brown 2015). This narrower view reduces politics to the overt (often democratic) governance of science and technology. It also obscures attention to more subtle operations of power within the epistemic cultures of scientific practice.

7 In this sense, we are knowers, too, and our work – including this book – is the product of a scientific inequality formation, reflecting our positions in the academy and our professional networks and interests. As such, we understand our own "academic" or "intellectual" goals to mark out essentially normative positions supported by claims generated by us and allied colleagues that are inherently situated and thus unavoidably partial.

2 Profitable Knowledge

1 "Whatever we think of alchemy today … many people accepted the basic principles of alchemy in early Modern Europe, even the transmutation of metals and could point to religious and natural philosophical justifications for their belief. [From this perspective,] successful alchemical works were rare and wondrous indeed, but not impossible" (Nummedal 2007: 16).

2 For Patel and Moore (2017: 22), cheapness is an analytical heuristic. They write, "Cheap is not the same as low cost, although that's part of it. Cheap is a strategy, a practice, a violence that mobilizes all kinds of work – human and animal, botanical and geological – with as little compensation as possible."

3 This section draws on historical information from books by Katy Barrett (2022); Jim Bennett (2017a and 2017b); and Dava Sobel (1995).

4 Sailors working on these ships were very specifically involved in the production of a commodity – the "slave" (Brown 2006; referenced in Rediker 2007).

5 Although we do not engage this work here, the emerging historical literature on racial capitalism, beginning with Cedric Robinson's classic study *Black Marxism* (1983), is an important resource for those interested in racialization and scientific inequality formation during this time period.

6 Schnaiberg did not theorize gender inequality as a structural feature of science.

7 After serving for one year, Nelson was replaced in 2022 by the new OSTP Director, Dr. Arati Prabhakar, who holds a Ph.D. in Applied Physics from Cal Tech. The only other impact science researcher to fill this post – albeit also in a temporary capacity as Acting Director (for 2001), was Rosina Bierbaum, a University of Michigan ecologist.

8 This section draws heavily on these two accounts.

9 As catalogued by Hess, the main positions of the debate are articulated in Callon and Law (1982); MacKenzie (1981), (1984); Woolgar (1981a and 1981b); and Yearley (1982).

10 As we describe in Chapter 4, recent shifts in scientific inequality formation within STS have come from a rising activist wing that promotes community-based research and political intervention and from the influence of intersectionalist, post-colonialist, Indigenous, and critical gender and critical race frameworks in STS scholarship.

11 While not an exhaustive list, these include: Harry Collins' (1992[1985]) "core-set" model; the social worlds approach developed by Adele Clarke, Joan Fujimura, and Susan Leigh Star (Clarke and Star 2008); Andrew Pickering's (1995) "mangle of practice"; the "social construction of technology" (SCOT) framework of Weibe Bijker, Thomas Hughes, and Trevor Pinch (1989); and, more recently, various takes on "performativity theory" (Callon 2007).

12 http://foundaryri.com/our-team.

13 Louis Berger Group, Inc. "Phase II Site Assessment Report, Foundry Apartments Property," Rhode Island Department of Environmental Management file box SR-28-0495 (January 23, 2003).

14 https://en.wikipedia.org/wiki/Louis_Berger_Group.

3 Absent-Minded Science

1 Harvey is the wettest storm on record for a major U.S. city, with peak rainfall over the full four-day period totaling more than 80 inches in some neighborhoods. https://www.wunderground.com/cat6/harvey-houston-most-extreme-rains-ever-major-us-city.

2 Federal pollution reporting requirements for industrial releases remained in place throughout.

3 Structural functionalism is a theoretical framework that understands society as a complex system of interacting institutions whose smooth functioning provides stability and social order. The defining text is Talcott Parson's *The Structure of Social Action*, published in 1937. For a sociological assessment, see Camic (1989).

4 Our thanks for Meghan Kallman for pointing out the potential for misinterpretation here.

5 From Merton's perspective, one could attempt a history of the development of any discipline as a collective effort to account for missing knowledge about ignored topics, albeit without a crucial politics of knowledge that addresses mechanisms of systematic exclusion and erasure.

6 Schiebinger addresses the larger question of intentionally hidden medical knowledge among seventeenth-century slave and Creole populations in a different study, *The Secret Cures of Slaves* (2017). In her conclusion, Schiebinger identifies five "barriers" to the circulation of Indigenous and African knowledge in the colonial context: genocide, enslavement, language differences, secrecy, and prejudice (158–164).

7 For more on eighteenth-century abortifacients, including the savin tree and caraguata-acanga, see Schiebinger 2000 and Sponsel and Schiebinger (2011).

8 Important critiques of objectivity from STS scholars include Bloor (1991); Pinch (1981); Haraway (1988); Harding (2015); Latour (1996). Early critics dismissed Merton's theory of scientific norms as ideology (see Mitroff 1974; Mulkay 1976). More recently, renewed interest in the theory has led to critical extensions of the framework in different contexts (see, e.g. Kim and Kim 2018; Macfarlane and Cheng 2008; Ziman 2000).

9 All were early American Sociological Society presidents and founding Sociology Department chairs at their respective universities. See Simpson and Simpson 1994.

4 Challenging Scientific Inequality Formation

1 Specifically, *Nature*, *Science*, and *Proceedings of the National Academy of Sciences*.

2 Henceforward, different levels of physical and biological restrictions would apply to low-, moderate-, and high-risk experiments. Lower risk experiments could proceed with limited physical and biological controls, but scientists pursuing high-risk experiments would need to conduct them in specialized containment facilities and use specially engineered organisms that could not survive outside the lab.

3 These include the Baye-Dohl Act (1980) (specifically §200 and §207), the Stevenson Wydler Technology Innovation Act (1980) and the Cartagena Protocol on Biosafety (2000); see also Kleinman and Kinchy (2007) and Vallas, Kleinman and Biscotti (2011).

4 STS long ago discarded the notion of an idealized separation between science and politics, most (in)famously with Bruno Latour's observation that "science is politics pursued by other means" (Latour 1983: 168, 1988; Seguin and Vinck 2023).

5 Soon after the *Pagina 12* article was published, the study also appeared in a major international scientific journal, *Chemical Research in Toxicology* (Paganelli et al. 2010).

6 The Biology Division at Oak Ridge National Laboratory housed the EMS office and staff; other key agencies were the National Institute for Environmental Health Sciences, Agency for Toxic Substances and Disease Registry, and EPA's Office of Research and Development.

7 The "Delany Clause" was a 1958 amendment to the Food and Drug Administration Act that banned food additives found to be carcinogenic in human and animal systems.

8 The links between environmental discourse and a reform eugenics movement is described in Frickel (2004a).

9 The Frank R. Lautenberg Chemical Safety for the 21st Century Act, passed in June 2016. For background and assessment, see Denison (2017), McLean (2020), and Schmidt (2016).

10 The recent Supreme Court decision overturning the so-called "Chevron Doctrine," by which judges were required to defer to agency expert interpretations where ambiguities exist in statutory law, makes such abuses all the more likely today. See Furlow (2024).

11 An acronym for Registration, Evaluation, Authorisation, and Restriction of Chemicals, REACH entered into force of EU law in 2007.

12 Cat 1B mutagens are chemicals that test positive for mutagenicity in animal studies (European Chemicals Agency 2018).

13 Some SIMs nurture direct connections to social movements that may prove advantageous for diversifying faculty and student populations on campus and training students for movement-related activism and employment after college. Comparative studies that assess the relative impact of different types of SIMs are as yet undone.

14 Prominent examples from the U.S. include Cornell Lab of Ornithology's "Great Backyard Bird Count"(among others; https://www.birds.cornell.edu/ho me/engagement-in-science-and-nature/), and NASA's "Active Asteroids" – one of 29 current projects open to "anyone with a smartphone or laptop" (https://science.nasa.gov/citizen-science/).

15 The four critiques are summarized on pages 4–5, but further developed throughout the book.
16 For details, see https://www.breatheprovidence.com/.
17 In the book, Kimura uses the term "citizen science" synonymously with our understanding of community science.

Provocation: Toward a Deeply Adapted Science
1 https://dictionary.cambridge.org/us/dictionary/english/erosion.
2 In his essay titled "The World as Laboratory," Beck (1995) argued that the global impacts of nuclear, genetic, and chemical technologies have undermined sciences' ability to conduct controlled ecological experiments. Climate change exacerbates the problem, obliterating distinctions between society and nature and between science and politics.

References

Abbott, Andrew. 2001. *Chaos of Discipline*. Chicago: University of Chicago Press.

Adami, George. 1921. "The True Aristocracy." *The Scientific Monthly* 13(5), November: 420–434.

Albagli Sarita, Maria Lucia Maciel, and Alexandre Hannud Abdo, eds. 2015. *Open Science, Open Issues*. Rio de Janeiro: IBICT and Unirio. http://livroaberto.ibict.br/bitstream/1/1061/1/Open%20Science%20 open%20issues_Digital.pdf.

Alegria, Sharla and Enobong Hannah Branch. 2015. "Causes and Consequences of Inequality in the STEM Fields: Diversity and its Discontents." *International Journal Gender, Science, and Technology* 7: 321–342.

Alic, Margaret. 2022. "Cobb, W. Montague 1904–1990." Contemporary Black Biography. *Encyclopedia.com*. (28 Feb). https:// www.encyclopedia.com/history/historians-and-chronicles/histori ans-miscellaneous-biographies/william-montague-cobb#F.

Allen, Barbara L. 2003. *Uneasy Alchemy: Citizens and Experts in Louisiana's Chemical Corridor Disputes*. Cambridge: MIT Press.

Allen, Barbara L. 2004. "Shifting boundary work: Issues and tensions in environmental health science in the case of Grand Bois, Louisiana." *Science as Culture* 13(4): 429–448.

Allen, Barbara L. 2018. "Strongly Participatory Science and Knowledge Justice in an Environmentally Contested Region." *Science, Technology, & Human Values*, 43(6): 947–971. https://doi.org/10 .1177/0162243918758380.

Altieri, Miguel. A. 1995. *Agroecology: the Science of Sustainable Agriculture*. Boulder: Westview Press.

Arancibia, Florencia. 2013. "Challenging the Bioeconomy: The Dynamics of Collective Action in Argentina." *Technology in Society* 35(2): 72–92.

Arancibia, Florencia and Renata Motta. 2019. "Undone Science and Counter-Expertise: Fighting for Justice in an Argentine Community Contaminated by Pesticides." *Science as Culture* 28(3):277–302. doi: 10.1080/09505431.2018.1533936.

Arancibia, Florencia, Valeria Arza, Damián Verzeñassi, and Scott Frickel. 2022. "Building Participatory Knowledge Infrastructure Against the GMO Agribusiness Regime: The Case of Los Campamentos Sanitarios." *Citizen Science: Theory and Practice* 7(1): 17.

Arancibia, Florencia, Scott Frickel, and Alexandre Annud Hannod. 2024. "The Political Shaping of Argentine Pesticide Science: Evidence from Domain-Topic Model Analysis." Paper presented at the Annual Meetings of the Society for Social Studies of Science, Amsterdam, The Netherlands (July).

Arancio, Julieta, and Shannon Dosemagen. 2022. "Bringing Open Source to the Global Lab Bench." *Issues in Science and Technology* 38(2):18–20.

Arthur, Mikaila Mariel Lemonik. 2009. "Thinking Outside the Master's House: New Knowledge Movements and the Emergence of Academic Disciplines." *Social Movement Studies* 8(1): 73–87. doi: 10.1080/14742830802591176.

Asher, Kiran. 2009. *Black and Green: Afro-Colombians, Development and Nature in the Pacific Lowlands.* Durham: Duke University Press.

Auyero, Javier and Debora Alejandra Swistun. 2009. *Flammable: Environmental Suffering in an Argentine Shantytown.* Oxford: Oxford University Press.

Bangham, Jenny, Xan Chacko, and Judith Kaplan, eds. 2022. *Invisible Labour in Modern Science.* New York: Rowman & Littlefield.

Barinaga, Marcia. 2000. "Asilomar Revisited: Lessons for Today?" *Science* 287(5458): 1584–1585. doi: 10.1126/science.287.5458.1584.

Barrett, Katy. 2022. *Looking for Longitude: A Cultural History.* Liverpool: Liverpool University Press.

Baurick, Timothy, Lylla Younes, and Joan Meiners. 2019. "Welcome to 'Cancer Alley,' Where Toxic Air Is About to Get Worse." *The Times-Picayune, The Advocate,* and *ProPublica,* October 30. https://www.propublica.org/article/welcome-to-cancer-alley-where-toxic-air-is-about-to-get-worse.

Beck, Ulrich. 1992. *Risk Society: Towards a New Modernity.* Thousand Oaks: Sage.

Beck, Ulrich. 1995. *Ecological Enlightenment: Essays on the Politics of the Risk Society,* trans. Mark Ritter. Cambridge: Polity.

Bendell, Jem. 2018. "Deep Adaptation: A Map for Navigating Climate Tragedy." Institute for Leadership and Sustainability (IFLAS)

Occasional Papers Volume 2. University of Cumbria, Ambleside, U.K. (Unpublished); https://orcid.org/0000-0003-0765-4413.

Bendell, Jem and Rupert Read, eds. 2021. *Deep Adaptation: Navigating the Realities of Climate Chaos*. Cambridge: Polity.

Benjamin, Ruha. 2013. *The People's Science: Bodies and Rights on the Stem Cell Frontier*. Stanford: Stanford University Press.

Benjamin, Ruha. 2019. *Race after Technology: Abolitionist Tools for the New Jim Code*. Cambridge: Polity.

Bennett, Jim. 2017a. *Navigation: A Very Short Introduction*. London and New York: Oxford University Press.

Bennett, Jim. 2017b. "The Historian and the Longitude." *OUPblog* (March 24). https://blog.oup.com/2017/03/historian-longitude-john-harrison-navigation/.

Berg, Paul. 2008. "Asilomar 1975: DNA Modification Secured." *Science* 445: 290–291.

Berg, Paul et al. 1974. "Potential Biohazards of Recombinant DNA Molecules." *PNAS* 71(7): 2593–2594.

Berros, Marìa Valeria. 2024. "Stop Experimenting on Us! Judicial Stories of Pesticide Resistance in Argentina." *Global Environment* 17(2): 378–395.

Bijker, Wiebe. E., Thomas P. Hughes, and Trevor Pinch, eds. 1989. *The Social Construction of Technological Systems*. Cambridge: MIT Press.

Bliss, Catherine. 2012. *Race Decoded: The Genomic Fight for Social Justice*. Palo Alto: Stanford University Press.

Block, Fred. 2008. "Swimming Against the Current: The Rise of a Hidden Developmental State in the United States." *Politics & Society* 36(2):169–206. doi: 10.1177/0032329208318731.

Bloor, David, 1991. *Knowledge and Social Imagery* (2nd edn.). Chicago: University of Chicago Press.

Blume, Stuart S. 1974. *Toward a Political Sociology of Science*. New York: Free Press.

Bourdieu, Pierre. 1988. *Homo Academicus*. Cambridge: Polity.

Braudel, Fernand. 2001. *Memory and the Mediterranean*. New York: Vintage.

Braverman, Harry. 1974. *Labor and Monopoly Capital: The Degradation of Work in the Twentieth Century*. New York: Monthly Review Press.

Bridges, Khiara M. 2011. *Reproducing Race: An Ethnography of Pregnancy as a Site of Racialization*. Berkeley: University of California Press.

Brint, Steven and Jerome Karabel. 1991. *The Diverted Dream: Community Colleges and the Promise of Educational Opportunity in America, 1900–1985*. London: Oxford University Press.

Brown, Christopher Leslie. 2006. *Moral Capital: Foundations of British Abolitionism*. Chapel Hill: University of North Carolina Press.
Brown, Mark B. 2015. "Politicizing Science: Conceptions of Politics in Science and Technology Studies." *Social Studies of Science* 45(1): 3–30.
Brown, Phil and Edwin J. Mikkelsen. 1997. *No Safe Place: Toxic Waste, Leukemia, and Community Action*. Berkeley: University of California Press.
Brown, Phil. 2007. *Toxic Exposures: Contested Illnesses and the Environmental Health Movement*. New York: Columbia University Press.
Bud, Robert. 1994. *The Uses of Life: A History of Biotechnology*. Cambridge: Cambridge University Press.
Burawoy, Michael. 2004. "For Public Sociology." *American Sociological Review* 70, February: 4–28.
Callon, Michel. 2007. "What Does it Mean to Say that Economics is Performative?" In *Do Economists Make Markets? On the Performativity of Economics*, ed. MacKenzie, D., Munisia, F., and Siu, L., 311–357. Princeton: Princeton University Press.
Callon, Michel, and John Law. 1982. "On Interests and Their Transformation: Enrollment and Counter Enrollment." *Social Studies of Science* 12: 615–25.
Camic, Charles M. 1989. "Structure after 50 Years: The Anatomy of a Charter." *American Journal of Sociology* 95(1): 38–107.
Camic, Charles M. 1995. "Three Departments in Search of a Discipline: Localism and Interdisciplinary Interaction in American Sociology, 1890–1940." *Social Research* 62: 1003–1033.
Carson, Rachel. 1962. *Silent Spring*. New York: Fawcett Crest.
Cassegård, Carl. 2024. "Activism without Hope? Four Varieties of Postapocalyptic Environmentalism." *Environmental Politics* 33(3): 444–464. doi:10.1080/09644016.2023.2226022.
Cech, Erin A., Anneke Metz, Jessi L. Smith and Karen deVries. 2017. "Epistemological Dominance and Social Inequality: Experiences of Native American Science, Engineering, and Health Students." *Science, Technology, & Human Values* 42(5): 743–774.
Chorev, Nitsan. 2018. "Neoliberalism and Supra-national Institutions." In *The SAGE Handbook of Neoliberalism*, ed. Damien Cahill, Melinda Cooper, Martijn Konings, and David Primrose, 260–269. London: Sage. https://doi.org/10.4135/9781526416001.
Clark, Adele and Susan Leigh Star. 2008. "The Social Worlds Framework: A Theory/Methods Package." In *The Handbook of Science and Technology Studies*, 3rd ed., ed. E. J. Hackett, 113–137. Cambridge: MIT Press.
Cobb, W. Montague. 1936. "Race and Runners." *Journal of Health and Physical Education* 1: 3–36.

Colborn, Theo, Dianne Dumanoski, and John P. Meyers. 1997. *Our Stolen Future: Are We Threatening Our Fertility, Intelligence, and Survival?* New York: Plume/Penguin Books.

Cole, Jonathan R. and Stephen Cole. 1973. *Social Stratification in Science.* Chicago: University of Chicago Press.

Cole, Luke W., and Sheila R. Foster. 2001. *From the Ground Up: Environmental Racism and the Rise of the Environmental Justice Movement.* New York: NYU Press.

Collins, Harry M. 1974. "The TEA Set: Tacit Knowledge and Scientific Networks." *Science Studies* 4(2): 165–85.

Collins, Harry M. 1992 [1985]. *Changing Order: Replication and Induction in Scientific Practice*, 2nd edn. Chicago: University of Chicago Press.

Creager, Angela. 2018. "Human Bodies as Chemical Sensors: A History of Biomonitoring for Environmental Health and Regulation." *Studies in History and Philosophy of Science* 70: 70–81.

Croissant, Jennifer L. 2014. "Agnotology: Ignorance and Absence or Towards a Sociology of Things That Aren't There." *Social Epistemology* 28(1): 4–25.

Crow, James F. 1968. "Chemical Risk to Future Generations." *Scientist and Citizen* 10:113–7.

Daston, Lorraine J. and Peter Galison. 1992. "The Image of Objectivity." *Representations* (special issue, Seeing Science) 40(Autumn): 81–128.

Daston, Lorraine J. and Peter Galison. 2007. *Objectivity.* Cambridge: MIT Press.

Davies, Thom. 2018. "Toxic Space and Time: Slow Violence, Necropolitics, and Petrochemical Pollution." *Annals of the American Association of Geographers* 108 (6): 1537–53. doi: 10.1080/24694452.2018.1470924.

De Châtel, Francesca. 2014. "The Role of Drought and Climate Change in the Syrian Uprising: Untangling the Triggers of the Revolution," *Middle Eastern Studies*, 50 (4), 521–535. doi: 10.1080/00263206.2013.850076.

Dearden, Jason and Michael Biesecker. 2017. "Toxic Waste Sites Flooded in Houston Area." *AP News*, September 3. https://apnews.com/27796dd13b9549b0ac76aded58a15122.

Delborne, Jason A. 2008. "Transgenes and Transgressions: Scientific Dissent as Heterogeneous Practice." *Social Studies of Science* 38(4): 509–541. doi:10.1177/0306312708089716.

Delborne, Jason A., Dresden Hasala, Aubrey Wigner, Abby Kinchy. 2020. "Dueling Metaphors, Fueling Futures: 'Bridge Fuel' Visions of Coal and Natural Gas in the United States." *Energy Research & Social Science* 61:101350. https://doi.org/10.1016/j.erss.2019.101350.

Denison, Richard. 2017. "A Primer on the New Toxic Substances Control Act (TSCA) and What Led to It." *Environmental Defense Fund*, April. www.edf.org/sites/default/files/denison-primer-on-lautenberg-act.pdf.

Denison, Richard. 2018. "EPA Rams Through its Reckless Review Scheme for New Chemicals under TSCA, Your Health be Damned." *EDF Health Blog*, August 1. https://blogs.edf.org/health/2018/08/01/epa-rams-through-its-reckless-review-scheme-for-new-chemicals-under-tsca-your-health-be-damned/.

Doing, Park. 2004. "Lab Hands and the 'Scarlet O': Epistemic Politics and (Scientific) Labor." *Social Studies of Science* 34 (3): 299–323.

Dosemagen, Shannon and Alison J. Parker. 2019. "Citizen Science Across a Spectrum: Broadening the Impact of Citizen Science and Community Science." *Science and Technology Studies* 32(2): 24–33.

Dunn, Richard S. 2007. "The Demographic Contrast between Slave Life in Jamaica and Virginia, 1760–1865." *Proceedings of the American Philosophical Society* 151(1): 43–60.

Duster, Troy. 1982. "Social Implications of the New Genetics Technology." In *Human Genetic Engineering*, Congressional Record, Sub-Committee on Investigations and Oversight of the Committee on Science and Technology, U.S. House of Representatives, November, 476–500.

Duster, Troy. 1991. "Assessing the Quality of Life: Genetic Screening, Medical Science and the Backdoor to Medical Eugenics." *Chronicle of Social Action* 4 (1).

Eastman, David A., Andrea Hartwig, Diana Anderson, Wagida A. Anwar, Michael C. Cimino, Ivan Dobrev, George R. Douglas, Takehiko Nohmi, David H. Phillips and Carolyn Vickers. 2009. "Mutagenicity Testing for Chemical Risk Assessment: Update of the WHO/IPCS Harmonized Scheme." *Mutagenesis* 24(4): 341–349.

Easton, David. 1965. *A Framework for Political Analysis*. Englewood Cliffs: Prentice-Hall.

Education Dynamics. 2023. *Survey of the Higher Education Landscape Report 2023*. https://insights.educationdynamics.com/higher-ed-landscape-report-content-2023.html?utm_source=meltwater&utm_medium=release&utm_campaign=landscape-2023.

Egert, Philip R. 2013. "A Conversation with David Hess about 'Neoliberalism and the History of STS Theory.'" *Social Epistemology Review and Reply Collective* 2 (11): 7–12.

Elliott, Rebecca. 2018. "The Sociology of Climate Change as a Sociology of Loss." *European Journal of Sociology* 59(3): 301–337.

Emigh, Rebecca Jean, Dylan Riley, and Patricia Ahmed. 2016. *Changes in Censuses from Imperialist to Welfare States*. London: Palgrave Macmillan.

Environmental Defense Fund. 1997. *Toxic Ignorance: The Continuing Absence of Basic Health Testing for Top-Selling Chemicals in the United States*. New York: Environmental Defense Fund.

Epps, Charles H. Jr, Davis G. Johnson, and Audrey L. Vaughan. 1993. "Black Medical Pioneers: African-American 'Firsts' in Academic and Organized Medicine." *Journal of the National Medical Association* 85(10): 777–796.

Epstein, Steven. 2007. *Inclusion: The Politics of Difference in Medical Research*. Chicago: University of Chicago Press.

Etzkowitz, Henry. 2008. *The Triple Helix: University–Industry–Government Innovation in Action*. London: Routledge.

European Chemicals Agency. 2018. *Background Document on Germ Cell Mutagenicity*. MSC-RAC Joint Workshop (11–12 October), Helsinki, Finland.

Eyal, Gil. 2019. *The Crisis of Expertise*. Cambridge: Polity.

Farber, Kathryn, Meg Fay, Grace Berg, Vivien Chen, Wendell Walters, and Meredith Hastings. 2023. "Unveiling the Local Impact of Wildfire Pollution Events Using the Breathe Providence Network." The Leadership Alliance Digital Library, Breathe Providence Digital Collection, Summer Research Symposium. Brown Digital Repository. Brown University Library. https://doi.org/10.26300/3grb-3c85.

Fay, Meg. 2023. "Fine Particulate Matter in Providence: Siting, Calibration, and Analysis of the Breathe Providence Hyperlocal Air Monitoring Network." Earth, Environmental and Planetary Sciences Theses and Dissertations, Institute at Brown for Environment & Society Theses and Dissertations, Breathe Providence Digital Collection. Brown Digital Repository. Brown University Library. https://doi.org/10.26300/k26t-kh39.

Fernández-Llamazares, Álvaro, María Garteizgogeascoa, Niladri Basu, Eduardo Sonnewend Brondizio, Mar Cabeza, Joan Martínez-Alier, Pamela McElwee, Victoria Reyes-García. 2020. "A State-of-the-Art Review of Indigenous Peoples and Environmental Pollution." *Integrated Environmental Assessment and Management* 16(3): 324–341.

Felter, Susan P., Virunya S. Bhat, Philip A. Botham, David A. Bussard, Warren Casey, A. Wallace Hayes, Gina M. Hilton, Kelly A. Magurany, Ursula G. Sauer & Edward V. Ohanian. 2021. "Assessing Chemical Carcinogenicity: Hazard Identification, Classification, and Risk Assessment. Insight from a Toxicology Forum State-of-the-Science Workshop." *Critical Reviews in Toxicology* 51:8, pages 653–694.

Field, Kelly. 2016. "For Native Students, A Deepening Divide." *Chronicle of Higher Education*, July 26. https://www.chronicle.com/article/for-native-students-a-deepening-divide/.

Firestein, Stuart. 2012. *Ignorance: How it Drives Science.* Oxford: Oxford University Press.

Fleurbaey, Marc. 2014. "The Facets of Exploitation." *Journal of Theoretical Politics* 26(4): 653–676.

Fox, Mary Frank, Kjersten Bunker Whittington, and Marcela Linková. 2017. "Gender, (In)equity, and the Scientific Workforce." In *The Handbook of Science and Technology Studies*, ed. Ulrike Felt, Rayvon Fouché, Clark Miller and Laurel Smith-Doerr, 701–732. Cambridge: MIT Press.

Franklin, Sarah. 2007. *Dolly Mixtures: The Remaking of Genealogy.* Durham: Duke University Press.

Fredrickson, Donald S. 1991. "Asilomar and Recombinant DNA: The End of the Beginning." In *Biomedical Politics*, ed. Kathi E. Hanna, 258–298. Committee to Study Decision Making, Division of Health Sciences Policy. Washington, DC: National Academies Press.

Frickel, Scott. 2004a. *Chemical Consequences: Environmental Mutagens, Scientist Activism, and the Rise of Genetic Toxicology.* New Brunswick: Rutgers University Press.

Frickel, Scott. 2004b. "Building an Interdiscipline: Collective Action Framing and the Rise of Genetic Toxicology." *Social Problems* 5(2): 269–287.

Frickel, Scott and Neil Gross. 2005. "A General Theory of Scientific/ Intellectual Movements." *American Sociological Review* 70(2): 204–232.

Frickel, Scott and Kelly Moore, eds. 2006a. *The New Political Sociology of Science: Institutions, Networks, and Power.* Madison: University of Wisconsin Press.

Frickel, Scott and Kelly Moore. 2006b. "Challenges and Prospects for a New Political Sociology of Science." In *The New Political Sociology of Science: Institutions, Networks, and Power*, ed. Scott Frickel and Kelly Moore, 3–31. Madison: University of Wisconsin Press.

Frickel, Scott, Sahra Gibbon, Jeff Howard, Joanna Kempner, Gwen Ottinger, and David Hess. 2010. "Undone Science: Charting Social Movement and Civil Society Challenges to Research Agenda Setting." *Science, Technology & Human Values* 35(4): 444–473.

Frickel, Scott. 2014a. "Absences: Methodological Note about Nothing, in Particular." *Social Epistemology* 28(1): 86–95. https://doi.org/10.1080/02691728.2013.862881.

Frickel, Scott. 2014b. "Not Here and Everywhere: The Non-Production of Knowledge." In *Routledge Handbook of Science, Technology and Society*, ed. Daniel Lee Kleinman and Kelly Moore, 256–269. London: Routledge.

Frickel, Scott and David J. Hess, eds. 2014. *Fields of Knowledge: Science, Politics, and Publics in the Neoliberal Age.* London: Emerald Press.

Frickel, Scott, Rebekah Torcasso, and Annika Anderson. 2015. "The Organization of Expert Activism: Shadow Mobilization in Two Social Movements." *Mobilization: An International Quarterly* 21(3): 305–323.

Frickel, Scott and Florencia Arancibia. 2022. "Mobilizing Environmental Experts and Expertise." In *Routledge Handbook of Environmental Movements*, ed. Marco Guigni and Maria Grasso, 278–292. London: Routledge.

Frickel, Scott, Apollonya Porcelli, Amy Teller and Aaron Niznik. 2022. "Embodied, Embedded or Both? Investigating Experts and Expertise in Two Greater Boston Social Movements." *Social Movement Studies* 23(5): 589–606.

Frickel, Scott. 2023. "Parcel History: Stories of Late Industrialism." Unpublished manuscript. Available on request.

Frickel, Scott and Fernando Tomás-Aponte. 2023. "Science Activism is Surging – Which Marks a Culture Shift Among Scientists." *The Conversation*, July 6. https://theconversation.com/science-activism-is -surging-which-marks-a-culture-shift-among-scientists-207454.

Friedberg, Errol. 2007. "A Brief History of the DNA Repair Field." *Cell Research* 18: 3–7. https://doi.org/10.1038/cr.2007.113

Fujimura, Joan H. and Ramya Rajagopalan. 2011. "Different Differences: The Use of 'Genetic Ancestry' Versus Race in Biomedical Human Genetic Research." *Social Studies of Science* 41(1): 5–30.

Fuller, Steve. 2000. *Thomas Kuhn: A Philosophical History for Our Times*. Chicago: University of Chicago Press.

Furlow, Bryant. 2024. "U.S. Supreme Court Overturns Chevron Doctrine." *The Lancet Respiratory Medicine* 12(9): e56. https:// doi.org/10.1016/S2213-2600(24)00224-8.

Future of Life Institute. 2023. "Pause Giant AI Experiments: An Open Letter." *Future of Life Institute*, March 22. https://futureoflife.org /open-letter/pause-giant-ai-experiments/.

Gendreau, Adam, Grace Berg, Wendell Walters, and Meredith Hastings. 2023. "Investigating Potential Causes of Intra-Urban Air Quality Variation via The Breathe Providence Network." Breathe Providence Digital Collection, Summer Research Symposium. Brown Digital Repository. Brown University Library. https://doi.org/10.26300/ gd30-gm78.

Gendreau, Adam Kyle. 2024. "Investigating the Role of the Transportation Sector on Providence's Intra-Urban Air Quality Trends: An Analysis of Local Impacts from Bus Routes, Shipping Corridors, and Freeways via the Breathe Providence Network and Related Policy Solutions." Breathe Providence Digital Collection, Institute at Brown for Environment & Society Theses and

Dissertations, Urban Studies Theses and Dissertations. Brown Digital Repository. Brown University Library. https://doi.org/10.26300/vat1-pd44.

Gerstle, Gary. 2022. *The Rise and Fall of the Neoliberal Order: America and the World in the Free Market Era.* London: Oxford University Press.

Gieryn, Thomas F. 1999. *Cultural Boundaries of Science: Credibility on the Line.* Chicago: University of Chicago Press.

Gould, Kenneth A. 2015. "Slowing the Nanotechnology Treadmill: Impact Science Versus Production Science for Sustainable Technological Development." *Environmental Sociology* 1(3): 143–151.

Gould, Stephen Jay. 1981. *The Mismeasure of Man.* New York: W.W. Norton.

Griego, Angel L., Aaron B. Flores, Timothy W. Collins, and Sara E. Grineski. 2020. "Social vulnerability, disaster assistance, and recovery: A population-based study of Hurricane Harvey in Greater Houston, Texas." *International Journal of Disaster Risk Reduction* 51: 101766. https://doi.org/10.1016/j.ijdrr.2020.101766.

Gross, Matthias. 2007. "The Unknown in Process: Dynamic Connections of Ignorance, Non-Knowledge and Related Concepts." *Critical Sociology* 55(5): 742–759.

Gross, Mathias and Linsey McGoey, eds. 2015. *Routledge International Handbook of Ignorance Studies.* London and New York: Routledge.

Guston, David H. 2005. "On Consensus and Voting in Science: From Asilomar to the National Toxicology Program." In *The New Political Sociology of Science: Institutions, Networks and Power*, ed. Scott Frickel and Kelly Moore, 378–405. Madison: University of Wisconsin Press.

Habermas, Jürgen. 1973. *Legitimation Crisis*, trans. by Thomas MacCarthy. Boston: Beacon Press.

Haerpfer, Christian, Ronald Inglehart, Alejandro Moreno, Christian Welzel, Kseniya Kizilova, Jaime Diez-Medrano, Marta Lagos, Pippa Norris, Eduard Ponarin, and Bi Puranen, eds. 2022. *World Values Survey: Round Seven – Country-Pooled Datafile Version 6.0.* Madrid, Spain, and Vienna, Austria: JD Systems Institute and WVSA Secretariat. https://doi.org/10.14281/18241.24.

Hagstrom, Warren O. 1965. *The Scientific Community.* New York: Basic Books.

Hammonds, Evelyn M. and Rebecca M. Herzig. 2009. *The Nature of Difference: the Sciences of Race from Jefferson to Genomics.* Cambridge: MIT Press.

Hamrai, Aimi. 2017. *Building Access: Universal Design and the Politics of Disability.* Minneapolis: University of Minesota Press.

Haraway, Donna.1988. "Situated Knowledges: The Science Question in Feminism and the Privilege of Partial Perspective." *Feminist Studies* 14 (3): 575–599.

Haraway, Donna. 1989. *Primate Visions: Gender, Race, and Nature in the World of Modern Science.* New York: Routledge.

Harding, Sandra. 1991. *Whose Science? Whose Knowledge? Thinking from Women's Lives.* Ithaca: Cornell University Press.

Harding, Sandra, ed. 1991. *The "Racial" Economy of Science: Toward a Democratic Future.* Bloomington: Indiana University Press.

Harding, Sandra G. 1998. *Is Science Multicultural? Postcolonialisms, Feminisms, and Epistemologies.* Bloomington: Indiana University Press.

Harding, Sandra J. 2002. "Must the Advance of Science Advance Global Inequality?" *International Studies Review* 4(2): 87–105.

Harding, Sandra J. 2006. "Two Influential Theories of Ignorance and Philosophy's Interests in Ignoring Them." *Hypatia* 21(3): 20–36.

Harding, Sandra J. 2015. *Objectivity and Diversity: Another Logic of Scientific Research.* Chicago: University of Chicago Press.

Hastings, Meredith. 2024. *Affidavit of Meredith Hastings. State of Rhode Island v. Rhode Island Recycled Metals, LLC.* Superior Court of Rhode Island, July 11.

Hatch, Anthony Ryan. 2016. *Blood Sugar: Racial Pharmacology and Food Justice in Black America.* Minneapolis: University of Minnesota Press.

Hatch, Anthony Ryan. 2022. "The Data will Not Save Us: Afropessimism and Racial Antimatter in the COVID-19 Pandemic." *Big Data & Society* 9(1): 1–13. https://doi.org/10.1177/20539517211067948.

Hayden, Cori. 2004. *When Nature Goes Public: The Making and Unmaking of Bioprospecting in Mexico.* Princeton: Princeton University Press.

Helm, Sabrina, Joya Kemper, Samantha White, and David Dean. 2024. "Exploring Climate-Reproductive Concern: Factors Influencing Hesitancy towards Parenthood in the Context of the Climate Crisis." *Environmental Sociology*: 1–15.

Henry, Emmanuel. 2017. *Ignorance Scientifique et Inaction Publique. Les Politiques de Santé au Travail.* Paris: Presses de Sciences Po.

Hermanowicz, Joseph C. 1998. *The Stars are Not Enough: Scientists – Their Passions and Professions.* Chicago: University of Chicago Press.

Hernández Vidal, Nathalia. 2022. "Pedagogies for Seed Sovereignty in Colombia: Epistemic, Territorial, and Gendered Dimensions." *Agriculture and Human Values* 39: 1217–1229. https://doi.org/10.1007/s10460-022-10310-9.

Hernández Vidal, Nathalia, and Kelly Moore. 2022. "Seed Schools in Colombia and the Generative Character of Sociotechnical Dissent." *Engaging Science, Technology, and Society* 8(1): 171–188. https://doi.org/10.17351/ests2022.487.

Hess, David J. 2013. "Neoliberalism and the History of STS Theory: Toward a Reflexive Sociology." *Social Epistemology* 27 (2): 177–193.

Hess, David J. 2016. *Undone Science: Social Movements, Mobilized Publics, and Industrial Transitions.* Cambridge: MIT Press.

Hess, David J. and Scott Frickel. 2014. "Introduction: Fields of Knowledge and Theory Traditions in the Sociology of Science." *Political Power and Social Theory* 27: 1–30.

Hess, David J., Sulfikar Amir, Scott Frickel, Daniel Lee Kleinman, Kelly Moore, and Logan Williams. 2016. "Structural Inequality and the Politics of Science and Technology." In *Handbook of Science and Technology Studies*, ed. Ulrike Felt, Rayvon Fouché, Clark Miller, and Laurel Smith-Doerr, 319–348. London and New York: Sage.

Hindess, Barry and Paul Hirst. 1977. *Mode of Production and Social Formation.* London: MacMillan Press.

Hirschman, Daniel. 2021. "Rediscovering the 1%: Knowledge Infrastructures and the Stylized Facts of Inequality." *American Journal of Sociology* 127(3): 739–786.

Hoffman, Karen. 2013. "Taking Precaution out of Toxic Water Pollutants Policy." *Science, Technology & Human Values* 38(6): 829–850.

Hoffman, Steve G. 2017. "Managing Ambiguities at the Edge of Knowledge: Research Strategy and Artificial Intelligence Labs in an Era of Academic Capitalism." *Science, Technology and Human Values* 42(4): 703–740.

Hoffman, Steve G. 2021. "A Story of Nimble Knowledge Production in an Era of Academic Capitalism." *Theory and Society* 50: 541–575.

Howell, Junia and James R. Elliott. 2019. "Damages Done: The Longitudinal Impacts of Natural Hazards on Wealth Inequality in the United States." *Social Problems* 66(3): 448467.

International Monetary Fund. 2021. *World Economic Outlook, October 2021: Recovery During a Pandemic.* Washington, DC: IMF. https://www.imf.org/en/Publications/WEO/Issues/2021/10/12/world-economic-outlook-october-2021/#Chapters.

International Science Council. n.d. "Science in the Private Sector." Retrieved November 21, 2023. https://council.science/actionplan/science-private-sector/.

Ioannidis, John P.A., Kevin W. Boyack, and Richard Klavans. 2014. "Estimates of the Continuously Publishing Core in the Scientific Workforce." *PLOS ONE* 9(7): e101698. doi:10. 1371/journal.pone.0101698.

Irwin, Aisling. 2018. "Citizen Science Comes of Age." *Nature* 562 (7728): 480–482.

Jasanoff, Sheila. 1990. *The Fifth Branch: Science Advisors as Policymakers*. Cambridge: Harvard University Press.

Kadandale, Sowmya, Robert Marten, Sarah L. Dalglish, Dheepa Rajan and David B. Hipgrave. 2020. "Primary Health Care and the Climate Crisis." *Bulletin of the World Health Organization* 98(11): 818–820. doi: 10.2471/BLT.20.252882.

Kaltenbrunner, Walter, Kean Birch, and Maria Amuchastegui 2022. "Editorial Work and the Peer Review Economy of STS Journals." *Science, Technology, & Human Values* 47(4): 670–697.

Karaye, Ibraheem, Kahler W. Stone, Gaston A. Casillas, Galen Newman, and Jennifer A. Horney. 2019. "A Spatial Analysis of Possible Environmental Exposures in Recreational Areas Impacted by Hurricane Harvey Flooding, Harris County, Texas." *Environmental Management* 64:381–390. https://doi.org/10.1007/s00267-019-01204-4.

Kaye, Joel. 1998. *Economy and Nature in the Fourteenth Century: Money, Market Exchange, and the Emergence of Scientific Thought*. Cambridge: Cambridge University Press.

Kempner, Joanna, Jon F. Merz, and Charles L. Bosk. 2011. "Forbidden Knowledge: Public Controversy and the Production of Nonknowledge." *Sociological Forum* 26(3): 475–500.

Kim, Hyeon-Wook Kim and Zia Quresh. 2020. *Growth in a Time of Change: Global and Country Perspectives on a New Agenda*. Washington: Brookings Institution Press

Kim, So Young and Yoonhoo Kim. 2018. "The Ethos of Science and Its Correlates: An Empirical Analysis of Scientists' Endorsement of Mertonian Norms." *Science, Technology and Human Values* 23(1): 1–24. https://doi.org/10.1177/0971721817744438.

Kimmerer, Robin Wall. 2015. *Braiding Seetgrass: Indigenous Wisdom, Scientific Knowledge and the Teaching of Plants*. Milkweed Editions.

Kimura, Aya H. 2016. *Radiation Brain Moms and Citizen Scientists: The Gender Politics of Food Contamination after Fukushima*. Durham: Duke University Press.

Kimura, Aya and Abby Kinchy. 2019. *Science by the People: Participation, Power, and the Politics of Environmental Knowledge*. New Brunswick: Rutgers University Press.

Kinchy, Abby. 2012. *Science, Seeds and Struggle: The Global Politics of Transgenic Crops*. Cambridge: MIT Press.

Kinchy, Abby J. and Daniel Lee Kleinman. 2005. "Democratizing Science, Debating Values," *Dissent*, Summer: 54–62.

Klein, Naomi. 2007. *The Shock Doctrine: The Rise of Disaster Capitalism*. New York: Knopf.

Kleinman, Daniel Lee. 2003. *Impure Cultures: University Biology and the World of Commerce*. Madison: University of Wisconsin Press.

Kleinman, Daniel Lee and Abby J. Kinchy. 2007. "Against the Neoliberal Steamroller? The Biosafety Protocol and the Social Regulation of Agricultural Biotechnologies." *Agriculture and Human Values* 24: 195–206.

Kleinman, Daniel Lee. 2010. "Science and Enterprise." *Science as Culture* 19 (3): 381–386.

Knorr Cetina, Karin. 1999. *Epistemic Cultures: How the Sciences Make Knowledge*. Cambridge: Harvard University Press.

Kohler, Robert E. 1990. "The Ph.D. Machine: Building on the Collegiate Base." *Isis* 81: 638–662.

Kolopenuk, Jessica. 2020. "Miskâsowin: Indigenous Science, Technology, and Society." *Genealogy* 4(1): 21. https://doi.org/10.3390/genealogy4010021.

Kousky, Carolyn. 2019. "The Role of Natural Disaster Insurance in Recovery and Risk Reduction." *Annual Review of Resources and Environment* 11: 399–418. https://doi.org/10.1146/annurev-resource-100518-094028.

Krimsky, Sheldon. 1982. *Genetic Alchemy: The Social History of the Recombinant DNA Controversy*. Cambridge: MIT Press.

Kuchinskaya, Olga. 2014. *The Politics of Invisibility: Public Knowledge about Radiation Health Effects after Chernobyl*. Cambridge: MIT Press.

Kuhn, Thomas. 1962. *The Structure of Scientific Revolutions*. Chicago: University of Chicago Press.

Kumeh, Titania. 2010 "Kettleman City's Toxic Web." *Mother Jones* (July/August).

Latour, Bruno and Steve Woolgar. 1979. *Laboratory Life: The Construction of Scientific Facts*. Princeton: Princeton University Press.

Latour, Bruno. 1983. "Give me a Laboratory and I Will Raise the World." In *Science Observed: Perspectives on the Social Study of Science*, ed. Karin Knorr-Cetina and Michael Mulkay, 141–169. Los Angeles: Sage.

Latour, Bruno. 1988. *The Pasteurization of France/Irreductions*. Cambridge: Harvard University Press.

Latour, Bruno. 1992. "Where are the Missing Masses? The Sociology of a Few Mundane Artifacts." In *Shaping Technology/Building Society: Studies in Sociotechnical Change*, ed. Wiebe E. Bijker and John Law, 225–258. Cambridge: MIT Press.

Latour, Bruno. 1996. "On Interobjectivity." *Mind, Culture, and Activity: An International Journal* 3: 228–245.

Latour, Bruno. 2018. *Down to Earth*. Cambridge: Polity.

Lave, Rebecca, Phillip Mirowski, and Samuel Randalls. 2010.

"Introduction: STS and Neoliberal Science." *Social Studies of Science* 40(5): 659–675.

Lavin, Nancy. 2024. "R.I. Superior Court Judge Shuts Down Providence Scrapyard Pending Fire Prevention Review." *Rhode Island Current*, July 12. https://rhodeislandcurrent.com/2024/07/12/r-i-superior-court-judge-shuts-down-providence-scrapyard-pending-fire-prevention-review/.

Lederberg, Joshua. 1997. "Some Early Stirrings (1950 ff) of Concern about Environmental Mutagens." *Environmental and Molecular Mutagenesis* 30: 3–10.

Legator, Marvin. 1970. "Chemical Mutagenesis Comes of Age: Environmental Implications." *Journal of Heredity* 61: 239–242.

Leguizamón, Amalia. 2014. "Modifying Argentina: GM Soy and Socio-environmental Change." *Geoforum*, 53:149–160. doi: 10.1016/j.geoforum.2013.04.001.

Lerner, Steve. 2005. *Diamond: A Struggle for Environmental Justice in Louisiana's Chemical Corridor.* Cambridge: MIT Press.

Lerner, Steve. 2010. *Sacrifice Zones The Front Lines of Toxic Chemical Exposure in the United States.* Cambridge: MIT Press.

Leslie, Jacque. 2010. "What's Killing the Babies of Kettleman City?" *Mother Jones* (July/August).

Lewontin, Richard, Leon Kamin and Steven Rose. 1984. *Not in Our Genes: Biology, Ideology and Human Nature.* New York: Pantheon Books.

Liboiron, Max. 2021. *Pollution is Colonization.* Durham: Duke University Press.

Lok, Corie. 2016. "Science's 1%: How Income Inequality is Getting Worse in Research." *Nature* 537: 471–473. doi:10.1038/537471a.

MacKenzie, Donald. 1978. "Statistical Theory and Social Interests: A Case Study." *Social Studies of Science* 8: 35–83.

MacKenzie, Donald. 1981. "Interests, Positivism, and History." *Social Studies of Science* 11(4): 498–501.

MacKenzie, Donald. 1984. "Reply to Yearley." *Studies in the History and Philosophy of Science* 15 (3): 251–259.

Macfarlane, B., and M. Cheng. 2008. "Communism, Universalism and Disinterestedness: Re-examining Contemporary Support Among Academics for Merton's Scientific Norms." *Journal of Academic Ethics* 6(1): 67–78.

MacFarlane, Key. 2019. "Time, Waste, and the City: The Rise of the Environmental Industry," *Antipode* 51(1): 225–247.

Malin, Stephanie A. and Meghan Elizabeth Kallman. 2022. *Building Something Better: Environmental Crises and the Promise of Community Change.* New Brunswick: Rutgers University Press.

Mannheim, Karl. 1936. *Ideology and Utopia: An Introduction to the Sociology of Knowledge*. London: Routledge & Keegan Paul.

Marsa, Linda. 2015. "The People's Scientist." *Discover Magazine*, October 1, 2015. https://www.discovermagazine.com/environment/the-peoples-scientist.

Martin, Brian. 2007. *Justice Ignited: The Dynamics of Backfire*. Lanham: Rowman & Littlefield.

Mascarenhas, Michael. 2018. "White Space and Dark Matter: Prying Open the Black Box of STS." *Science, Technology, & Human Values* 43(2): 151–70. https://www.jstor.org/stable/26580445.

McLean, Kevin. 2020. "Three Years After – Where Does Implementation of the Lautenberg Act Stand?" *Harvard Law School, Environmental and Energy Law Program*, February 26. https://eelp.law.harvard.edu/wp-content/uploads/McLean-TSCA.pdf.

McGoey, Linsey. 2019. *The Unknowers: How Strategic Ignorance Rules the World*. London: Zed Books.

Merelman, Richard M. 2000. "Technological Cultures and Liberal Democracy in the United States." *Science, Technology & Human Values* 25(2): 167–194.

Merton, Robert K. 1968. *Social Theory and Social Structure*. Boston: Free Press.

Merton, Robert K. 1973. *The Sociology of Science: Theoretical and Empirical Investigations*. Chicago: University of Chicago Press.

Merton, Robert K. 1987. "Three Fragments from a Sociologist's Notebooks: Establishing the Phenomenon, Specified Ignorance, and Strategic Research Materials." *Annual Review of Sociology* 13: 1–28.

Mills, Charles W. 2008. "White Ignorance." In *Agnotology: The Making and Unmaking of Ignorance* ed. Robert N. Proctor and Londa Scheibinger, 230–249. Palo Alto: Stanford University Press.

Mitroff, Ian. 1974. "Norms and Counter-norms in a Select Group of the Apollo Moon Scientists: A Case Study of the Ambivalence of Scientists." *American Sociological Review* 39(4): 579–595.

Mirowski, Phillip and Dieter Plehwe, eds. 2009. *The Road from Mont Pèlerin: The Making of the Neoliberal Thought Collective*. Cambridge: Harvard University Press.

Moore, Kelly, Daniel Kleinman, David Hess, and Scott Frickel. 2011. "Science and Neoliberal Globalization: A Political Sociological Approach." *Theory and Society* 40: 505–532.

Moore, Wilbert E, and Melvin M. Tumin. 1949. "Some Social Functions of Ignorance." *American Sociological Review* 14(6): 787–795. https://doi.org/10.2307/2086681.

Morris, Aldon D. 2016. *The Scholar Denied: W.E.B. Du Bois and the Birth of Modern Sociology*. Oakland: University of California Press.

Mulkay, Michael J. 1976. "Norms and Ideology in Science." *Social Science Information* 15(4/5): 637–656.

Mullen, Rick. 2020. "C&EN Talks with Wilma Subra." *Chemical and Engineering News* 98(3): 26–27.

Murphy, M. 2006. *Sick Building Syndrome and the Problem of Uncertainty: Environmental Politics, Technoscience, and Women Workers.* Durham: Duke University Press.

Murphy, M. 2015. "Troubling Transnational Itineraries of Care in Feminist Health Practices." *Social Studies of Science* 45(5): 717–737.

Murphy, M. 2017a. *The Economization of Life.* Durham: Duke University Press.

Murphy, M. 2017b. "Alterlife and Decolonial Chemical Relations." *Cultural Anthropology* 32(4): 494–503.

National Center for Education Statistics (NCES). 2022. *Digest of Education Statistics,* Table 330.10. https://nces.ed.gov/programs/digest/d22/tables/dt22_330.10.asp.

National Science Board (NSB). 2022a. *Science and Technology: Public Perceptions, Awareness, and Information Sources.* Alexandria: National Science Foundation. https://ncses.nsf.gov/pubs/nsb20227.

National Science Board, National Science Foundation. 2022b. *Research and Development: U.S. Trends and International Comparisons. Science and Engineering Indicators 2022.* NSB-2022-5. Alexandria: National Science Foundation. https://ncses.nsf.gov/pubs/nsb20225/.

Nelkin, Dorothy. 2001. "Beyond Risk. Reporting about Genetics in the Post-Asilomar Press." *Perspectives in Biology and Medicine* 44(2): 199–207. doi: 10.1353/pbm.2001.0032.

Nelson, Alondra. 2016. *The Social Life of DNA: Race, Reparations, and Reconciliation After the Genome.* Boston: Beacon Press.

Nelson, Donna, and Lynnette D. Madsen. 2018. "Representation of Native Americans in U.S. Science and Engineering Faculty." *MRS Bulletin* 43(5): 379–383. doi:10.1557/mrs.2018.108.

Nicholas, Thomas, Galen Hall and Colleen Schmidt. 2020. "The Faulty Science, Doomism, and Flawed Conclusions of 'Deep Adaptation.'" *openDemocracy,* July 14. https://www.opendemocracy.net/en/oureconomy/faulty-science-doomism-and-flawed-conclusions-deep-adaptation/.

Nickolai, Daniel H., Steve G. Hoffman, and Mary Nell Trautner. 2012. "Can a Knowledge Sanctuary also be an Economic Engine? The Marketization of Higher Education as Institutional Boundary Work." *Sociology Compass* 6(3): 205–218.

Noble, Safiyah Umoja. 2018. *Algorithms of Oppression: How Search Engines Reinforce Racism.* New York: New York University Press.

Norgaard, Kari Marie. 2019. *Salmon and Acorns Feed Our People: Colonialism, Nature and Social Action.* New Brunswick: Rutgers University Press.

Nsude, Chinedu C., Rebecca Loraamm. Joshua J. Wimhurst, God'sgift N. Chukwuonye, and Ramit Debnath. 2024. "Renewables but Unjust? Critical Restoration Geography as a Framework for Global Renewable Energy Injustice." *Energy Research & Social Science* 114: 103609.

Nummedal, Tara. 2007. *Alchemy and Authority in the Holy Roman Empire*. Chicago: University of Chicago Press.

Omi, Michael and Howard Winant. 1994. *Racial Formation in the United States: From the 1960s to the 1990s* (2nd edn.). New York and London: Routledge.

Ottinger, Gwen. 2010. "Buckets of Resistance: Standards and the Effectiveness of Citizen Science." *Science, Technology, and Human Values* 35 (2): 244–270.

Ottinger, Gwen. 2013a. *Refining Expertise: How Responsible Engineers Subvert Environmental Justice Challenges*. New York: New York University Press.

Ottinger, Gwen. 2013b. "The Winds of Change: Environmental Justice in Energy Transitions." *Science as Culture* 22(2): 222–229. doi: 10.1080/09505431.2013.786996.

Ottinger, Gwen. 2017. "Crowdsourcing Undone Science." *Engaging Science, Technology and Society* 3: 560–574.

Ottinger, Gwen and Elisa Sarantschin. 2017. "Exposing Infrastructure: How Activists and Experts Connect Ambient Air Monitoring and Environmental Health." *Environmental Sociology* 3(2): 155–165.

Organisation for Economic Co-operation and Development (OECD). 2021. *Education at a Glance 2021: OECD Indicators*. Paris: OECD Publishing. https://doi.org/10.1787/b35a14e5-en.

Paganelli, Alejandra, Victoria Gnazzo, Helena Acosta, Silvia L López, and Andrés E Carrasco. 2010. "Glyphosate-based Herbicides Produce Teratogenic Effects on Vertebrates by Impairing Retinoic Acid Signaling." *Chemical Research in Toxicology* 23(10): 1586–1595.

Pais, Jeremy and James R. Elliott. 2008. "Places as Recovery Machines: Vulnerability and Neighborhood Change after Major Hurricanes." *Social Forces* 86(4): 1415–1453.

Pan, Bowen, Yuan Wang, Timothy Logan, Jen-Shan Hsieh, Jonathan H. Jiang, Yixin Li, and Renyi Zhang. 2020. "Determinant Role of Aerosols from Industrial Sources in Hurricane Harvey's Catastrophe." *Geophysical Research Letters* 47(23). https://doi.org/10.1029/2020 GL090014.

Panofsky, Aaron and Catherine Bliss. 2017. "Ambiguity and Scientific Authority: Population Classification in Genomic Science." *American Sociological Review* 82(1): 59–87.

Pardy, Martina, Capucine Riom and Roman Hoffmann. 2024. "Climate

Impacts on Material Wealth Inequality: Global Evidence from a Subnational Dataset." *Geography and Environment Discussion Paper Series* (48). Department of Geography and Environment, The London School of Economics and Political Science, London, U.K.

Parks, Robbie M., G. Brooke Anderson, Rachel C. Nethery, Ana Navas-Acien, Francesca Dominici and Marianthi-Anna Kiourmourtzoglou. 2021. "Tropical Cyclone Exposure is Associated with Increased Hospitalization Rates in Older Adults." *Nature Communication* 12: 1545. https://doi.org/10.1038/s41467-021-21777-1.

Parthasarathy, Shobita. 2017. *Patent Politics: Life Forms, Markets, and the Public Interest in the United States and Europe.* Chicago: University of Chicago Press.

Patel, Raj, and Jason W. Moore. 2017. *A History of the World in Seven Cheap Things: A Guide to Capitalism, Nature, and the Future of the Planet.* Los Angeles: University of California Press.

Pellow, David N. 2000. "Environmental Inequality Formation: Toward a Theory of Environmental Injustice." *American Behavioral Scientist* 43(4): 581–601.

Pellow, David Naguib. 2002. *Garbage Wars: The Struggle for Environmental Justice in Chicago.* Cambridge: MIT Press.

Phillips, Ari. 2018. *Preparing for the Next Storm: Learning from the Man-Made Environmental Disasters that Followed Hurricane Harvey.* Washington: Environmental Integrity Project.

Pickering, Andrew, ed. 1992. *Science as Practice and Culture.* Chicago: University of Chicago Press.

Pickering, Andrew. 1995. *The Mangle of Practice: Time, Agency and Science.* Chicago: University of Chicago Press.

Pickett, Kate E., and Richard G. Wilkinson. 2015. "Income Inequality and Health: A Causal Review." *Social Science and Medicine* 128: 316–326. https://doi.org/10.1016/j.socscimed.2014.12.031.

Pinch, Trevor. 1981. "The Sun-Set: On the Presentation of Certainty in Scientific Life." *Social Studies of Science* 11: 131–158.

Pitts-Taylor, Victoria. 2019. "Neurobiologically Poor? Brain Phenotypes, Inequality, and Biosocial Determinism." *Science, Technology & Human Values* 44(4): 660–685.

Porcelli, Apollonya, Scott Frickel and Aaron Niznik. 2022. "Long After 'People Before Highways': Social Movements and Knowledge Politics in Greater Boston, 1960–2016," *Social Problems* 70(3): 791–808. https://doi.org/10.1093/socpro/spac048.

Postsecondary National Policy Institute. 2021. *Native American Students in Higher Education Fact Sheet.* https://pnpi.org/native-american-students/.

Preda, Alex. 2023. *The Spectacle of Expertise: Why Financial Analysts Perform in the Media.* New York: Columbia University Press.

Preston, Julian R. and George R. Hoffman. 2001. "Genetic Toxicology." In *Casarett and Doull's Toxicology: The Basic Science of Poisons*, ed. C.D. Klaassen, 321–50. New York: McGraw-Hill.

Proctor, Robert N. and Londa Schiebinger, eds. 2008. *Agnotology: The Making and Unmaking of Ignorance*. Stanford: Stanford University Press.

Prüss-Üstün, Annette, J. Wolf, Carlos F. Corvalán, R. Bos, and María Purificación Neira. 2016. *Preventing Disease Through Healthy Environments: A Global Assessment of the Burden of Disease from Environmental Risks* (2nd edn.). Geneva: World Health Organization.

Puig de la Bellacasa, Maria. 2011. "Matters of Care in Technoscience: Assembling Neglected Things." *Social Studies of Science* 41(1): 85–106.

Rahman, Muhammad Habibur, Nejat Anbarci, and Mehmet A. Ulubaşoğlu. 2022. "'Storm Autocracies': Islands as Natural Experiments." *Journal of Development Economics* 159: 102982. https://doi.org/10.1016/j.jdeveco.2022.102982.

Reardon, Jenny. 2004. *Race to the Finish: Identity and Governance in the Age of Genomics*. Princeton: Princeton University Press.

Rediker, Marcus. 2007. *The Slave Ship: A Human History*. New York: Viking Penguin.

Richter, Lauren. 2017. "Constructing Insignificance: Critical Race Perspectives on Institutional Failure in Environmental Justice Communities." *Environmental Sociology* 4 (1): 107–21. doi:10.1080/23251042.2017.1410988.

Richter, Lauren, Alissa Cordner, Phil Brown. 2018. "Non-stick Science: Sixty Years of Research and (In)action on Fluorinated Compounds." *Social Studies of Science* 48(5): 691–714. doi:10.1177/0306312718799960.

Roberts, Timmons J. and Melissa M. Toffolon-Weiss. 2001. *Chronicles from the Environmental Justice Frontline*. Cambridge: Cambridge University Press.

Robinson, Cedric. 1983. *Black Marxism: The Making of the Black Radical Tradition*. London: Zed Books.

Rodríguez-Muñiz, Michael. 2015. "Intellectual Inheritances: Cultural Diagnostics and the State of Poverty Knowledge." *American Journal of Cultural Sociology* 3: 89–122. https://doi.org/10.1057/ajcs.2014.16.

Rohde, Joy. 2013. *Armed with Expertise: The Militarization of American Social Research During the Cold War*. Ithaca: Cornell University Press.

Rohde, Joy. 2017. "Pax Technologica: Computers, International Affairs, and Human Reason in the Cold War." *Isis* 108(4): 792–813.

Rose, Hilary. 1975. "The Social Determinants of Reproduction Science and Technology" in Karin Knorr and H. Strasser, eds. *Yearbook of the Sociology of Science*. Dordrecht: Reidel.

Rudy, Alan P., Dawn Coppin, Jason Konefal, Bradley T. Shaw, Toby Ten Eyck, Craig Harris and Lawrence Busch. 2007. *Universities in the Age of Corporate Science: The UC-Berkeley–Novartis Controversy*. Philadelphia: Temple University Press.

Rust, Susanne and Louis Sahagun. 2019. "Post-Hurricane Harvey, NASA tried to Fly a Pollution-Spotting Plane over Houston. The EPA said No." *Los Angeles Times*, March 5. https://www.latimes.com/local/california/la-me-nasa-jet-epa-hurricane-harvey-20190305-story.html.

Saforcada, Fernanda and Yamile Socolovsky. 2019. "Privatisation and Commercialisation of Universities in Latin America." *Education International*, November 19. https://issuu.com/educationinternational/docs/debate_sobre_la_privatizacion_english__digital_?fr=sZWM5ODQxMzY3Ng.

Saltmarsh, Chris. 2022. "Deep Adaptation – or Climate Justice?" *The Ecologist*, February 1. https://theecologist.org/2022/feb/01/deep-adaptation-or-climate-justice.

Sarewitz, Daniel. 1996. *Frontiers of Illusion: Science, Technology, and the Politics of Progress*. Philadelphia: Temple University Press.

Schiebinger, Londa. 1993. *Nature's Body: Gender in the Making of Modern Science*. Boston: Beacon Press.

Schiebinger, Londa. 2000. "Exotic Abortifacients: The Global Politics of Plants in the 18th Century." *Endeavour* 24(3): 117–121.

Schiebinger, Londa. 2004. *Plants and Empire: Colonial Bioprospecting in the Atlantic World*. Cambridge: Harvard University Press.

Schiebinger, Londa. 2008. "West Indian Abortifacients and the Making of Ignorance." In *Agnotology: The Making and Unmaking of Ignorance*, ed. Robert N. Proctor and Londa Scheibinger, 149–162. Palo Alto: Stanford University Press.

Schiebinger, Londa. 2017. *Secret Cures of Slaves: People, Plants, and Medicine in the Eighteenth-Century Atlantic World*. Stanford: Stanford University Press.

Schmalzer Sigrid, Daniel S. Chard, Alyssa Botelho. 2019. *Science for the People: Documents from America's Movement of Radical Scientists*. Amherst: University of Massachusetts Press.

Schmidt, Charles W. 2016. "TSCA 2.0: A New Era of Chemical Risk Management." *Environmental Health Perspectives* 124(10): 182–186.

Schnaiberg, Allan. 1977. "Obstacles to Environmental Research by Scientists and Technologists: A Social Structural Analysis." *Social Problems* 55(4): 501–520.

Schurman, Rachel and William A. Munro. 2013. *Fighting for the*

Future of Food: Activists versus Agribusiness in the Struggle over Biotechnology. Minneapolis: University of Minnesota Press.

Schweber, Libby. 2006. *Disciplining Statistics: Demography and Vital Statistics in France and England, 1830–1885.* Durham: Duke University Press.

Seguin, Eve and Dominique Vinck. 2023. "Introduction: *Science is Politics by Other Means* Revisited." *Perspectives on Science* 31(1): 1–8.

Selcer, Perrin. 2012. "Beyond the Cephalic Index: Negotiating Politics to Produce UNESCO's Scientific Statements on Race." *Current Anthropology* 53(S5): S173–S184.

Servigne, Pablo, and Raphaël Stevens. 2020. *How Everything Can Collapse: A Manual for Our Times*, trans. Andrew Brown. Cambridge: Polity.

Shapin, Steven. 1989. "The Invisible Technician." *Scientific American* 77(6): 554–563.

Shapin, Steven. 1994. *A Social History of Truth: Civility and Science in the Seventeenth Century.* Chicago: University of Chicago Press.

Shapin, Steven. 2008. *The Scientific Life: A Moral History of a Late Modern Vocation.* Chicago: University of Chicago Press.

Shapin, Steven. 2010. *Never Pure.* Baltimore: Johns Hopkins University Press.

Sheng, Xin, Carolyn Chisadza, Rangan Gupta, and Christian Pierdzioch. 2023. "Climate Shocks and Wealth Inequality in the U.K.: Evidence from Monthly Data." *Environmental Science and Pollution Research* 30: 77771–77783. https://doi.org/10.1007/s11356-023-27342-1.

Shi, Linda, Anjali Fisher, Rebecca M. Brenner, Amelia Greiner-Safi, Christine Shepard, and Jamie Vanucchi. 2022. "Equitable Buyouts? Learning from State, County, and Local Floodplain Management Programs." *Climatic Change* 174: 29. https://doi.org/10.1007/s105 84-022-03453-5.

Shim, Janet K. 2005. "Constructing Race Across the Science-Lay Divide: Racial Formation in the Epidemiology and Experience of Cardiovascular Disease." *Social Studies of Science* 35(3): 405–436.

Sica, Alan. 2007. "Defining Disciplinary Identity: The Historiography of U.S. Sociology." In *Sociology in America: A History*, ed. Craig Calhoun, 713–732. Chicago: University of Chicago Press.

Simpson, Ida Harper and Richard L. Simpson. 1994. "The Transformation of the American Sociological Association." *Sociological Forum* 9(2): 259–278.

Singer, Maxine F. 1977. "Scientists and the Control of Science." *New Scientist* 74(1056): 631–634.

Singer, Maxine F. 1979. "Spectacular Science and Ponderous Process." *Science* 203(4375): 9.

Slaughter, Sheila and Gary Rhoades. 2009. *Academic Capitalism: Markets, State, and Higher Education*. Baltimore: Johns Hopkins University Press.

Slota, Stephen C. and Geoffrey C. Bowker. 2016. "How Infrastructures Matter." In *The Handbook of Science and Technology Studies*, ed. Ulrike Felt, Rayvon Fouché, Clark A. Miller, and Laurel Smith-Doerr, 529–554. Cambridge: MIT Press.

Smith-Doerr, Laurel, Sharla Alegria, and Timothy Sacco. 2017. "How Diversity Matters in the U.S. Science and Engineering Workforce: A Critical Review Considering Teams, Fields, and Organizational Contexts." *Engaging Science, Technology and Society* 3: 139–153. https://estsjournal.org/index.php/ests/article/view/142.

Smith-Doerr, Lauren, Sharla Alegria, Kaye Husbands Fealing, Debra Fitzpatrick, and Donald Tomaskovic-Devey. 2019. "Gender Pay Gaps in U.S. Federal Science Agencies: An Organizational Approach." *American Journal of Sociology* 125(2): 534–576.

Smithson, Michael. 1989. *Ignorance and Uncertainty: Emerging Paradigms*. New York: Springer.

Sobel, Dava. 1995. *Longitude: The True Story of a Lone Genius Who Solved the Greatest Scientific Problem of His Time*. London: Walker and Company.

Sponsel, Alistair and Londa Schiebinger. 2011. "Selective Memory: An Interview with Londa Schiebinger." *Cabinet* 42 (Summer); https://cabinetmagazine.org/issues/42/sponsel_schiebinger.php.

Strathern, Marilyn (ed.). 2000. *Audit Cultures: Anthropological Studies in Accountability, Ethics, and the Academy*. London: Routledge.

Sullivan, Shannon, and Nancy Tuana, eds. 2007. *Race and Epistemologies of Ignorance*. Albany: SUNY Press.

Svampa, Maristella. 2015. "Commodities Consensus: Neoextractivism and Enclosure of the Commons in Latin America." *South Atlantic Quarterly* 114(1): 65–82.

TallBear, Kimberley. 2013. *Native American DNA*. Minneapolis: University of Minnesota Press.

TallBear, Kimberley. 2014. "The Emergence, Politics, and Marketplace of Native American DNA." In *The Routledge Handbook of Science, Technology and Society*, ed. Kelly Moore and Daniel L. Kleinman. London and New York: Routledge.

Taylor, Dorceta. 2014. *Toxic Communities: Environmental Racism, Industrial Pollution, and Residential Mobility*. New York: New York University Press.

Tilly, Charles. 1998. *Durable Inequality*. Chicago: University of Chicago Press.

Tuana, Nancy. 2004. "Coming to understand: Orgasm and the epistemology of ignorance." *Hypatia* 19(1): 194–232.

Tuck, Eve and K. Wayne Yang. 2014. "Unbecoming Claims: Pedagogies of Refusal in Qualitative Research." *Qualitative Inquiry* 20(6): 811–818.

Turchin, Peter. 2023. *End Times: Elites, Counter-Elites and the Path of Political Disintegration.* New York: Penguin.

United Nations Human Development Programme (UNHDP). 2022. *Human Development Report, 2021–22:Uncertain Times, Unsettled Lives: Shaping our Future in a Transforming World.* New York: United Nations. https://hdr.undp.org/content/human-development -report-2021-22.

Ureta, Sebastián. 2017. "Baselining Pollution: Producing 'Natural Soil' for an Environmental Risk Assessment Exercise in Chile." *Journal of Environmental Policy & Planning* 20(3): 342–355. https://doi.org/ 10.1080/1523908X.2017.1410430.

Ureta, Sebastián. 2021. "Ruination Science: Producing Knowledge from a Toxic World." *Science, Technology, & Human Values* 46(1): 29–52. https://doi.org/10.1177/0162243919900957.

Vallas, Stephen P. and Daniel Lee Kleinman. 2008. "Contradiction, Convergence and the Knowledge Economy: The Confluence of Academic and Commercial Biotechnology." *Socio-Economic Review* 6(2): 283–311.

Vallas, Steven P., Daniel Lee Kleinman, and Dina Biscotti. 2011. "Political Structures and the Making of U.S. Biotechnology." In *State of Innovation: The U.S. Government's Role in Technology Development*, ed. Fred L. Block and Matthew R. Keller, 57–76. London: Routledge.

van Bommel, Natascha and Johanna I. Höffken. 2023. "The Urgency of Climate Action and the Aim for Justice in Energy Transitions – Dynamics and Complexity." *Environmental Innovation and Societal Transitions* 48: 100763.

Vincent, Shirley, J. Timmons Roberts, and Stephen Mulkey. 2016. "Interdisciplinary Environmental and Sustainability Education: Islands of Progress in a Sea of Dysfunction." *Journal of Environmental Studies and Sciences* 6 (2): 418–424. doi:10.1007/s13412-015-0279-z.

Vogel, Sarah A. and Jody A. Roberts. 2011. "Why the Toxic Substances Control Act Needs an Overhaul, and How to Strengthen Oversight of Chemicals in the Interim." *Health Affairs* 30(5): 898–905.

Waidzunas, Tom J. 2013. "Intellectual Opportunity Structures and Science-targeted Activism: Influence of the Ex-Gay Movement on the Science of Sexual Orientation." *Mobilization* 18(1): 1–18.

Wassom, John. 1989. "Origins of Genetic Toxicology and the Environmental Mutagen Society." *Environmental and Molecular Mutagenesis* 14, Supplement 16: 1–6.

Watkins, Rachel J. 2007. "Knowledge from the Margins: W. Montague Cobb's Pioneering Research in Biocultural Anthropology." *American Anthropologist* 109(1): 86–196.

Weber, Max. 1949. *Max Weber on the Methodology of the Social Sciences*, trans. and ed. Edward A. Shils and Henry A. Finch. Glencoe: Free Press.

White, Richard. 1991. *It's Your Misfortune and None of My Own: A New History of the American West*. Norman: University of Oklahoma Press.

Whyte, Kyle Powys. 2019. "The Dakota Access Pipeline, Environmental Injustice, and U.S. Settler Colonialism." In *The Nature of Hope: Grassroots Organizing, Environmental Justice, and Political Change*, ed. Char Miller and Jeff Crane, 320–338. Boulder: University Press of Colorado.

Wickham, Grace M. and Thomas E. Shriver. 2021. "Emerging Contaminants, Coerced Ignorance and Environmental Health Concerns: The Case of Per- and Polyfluoroalkyl Substances (PFAS)." *Sociology of Health and Illness* 43(3):764–778. https://doi.org/10.11 11/1467-9566.13253.

Wing, Oliver E.J., William Lehman, Paul D. Bates, Christopher C. Sampson, Niall Quinn, Andrew M. Smith, Jeffrey C. Neal, Jeremy R. Porter, and Carolyn Kousky. 2022. "Inequitable Patterns of U.S. Flood Risk in the Anthropocene." *Nature Climate Change* 12: 156–162. https://doi.org/10.1038/s41558-021-01265-6.

Woolgar, Steve. 1981a. "Interests and Explanation in the Social Study of Science." *Social Studies of Science* 11(3): 365–394.

Woolgar, Steve. 1981b. "Critique and Criticism: Two Readings of Ethnomethodology." *Social Studies of Science* 11(4): 504–514.

Woolgar, Steve, Catelijne Coopmans, and Daniel Neyland. 2009. "Does STS Mean Business?" *Organization* 16(1): 5–30.

Woutersen, M., Beekman, M., Pronk, M.E.J., Muller, A., de Knecht, J.A., and Hakkert, B.C. 2018. "Does REACH Provide Sufficient Information to Regulate Mutagenic and Carcinogenic Substances?" *Human and Ecological Risk Assessment: An International Journal* 25(8): 1996–2016. https://doi.org/10.1080/10807039.2018.148 0351.

Yearley, Steven. 1982. "The Relationship between Epistemological and Sociological Cognitive Interests: Some Ambiguities Underlying the Use of Interest Theory in the Study of Scientific Knowledge." *Studies in the History and Philosophy of Science* 13(4): 353–388.

Young, Rachel and Solomon Hsiang. 2024. "Mortality Caused by Tropical Cyclones in the United States." *Nature* 635: 121–128. https://doi.org/10.1038/s41586-024-07945-5.

Zhang, Sarah. 2017. "A Chemical Plant Catches Fire after Harvey

Flooding." *The Atlantic*, August 31. https://www.theatlantic.com/science/archive/2017/08/harvey-flooding-explosion-petrochemicals/538560/.

Ziman, J. 2000. *Real Science: What It Is, and What It Means.* Cambridge: Cambridge University Press.

Zippel, Katherin. 2017. *Women in Global Science: Advancing Academic Careers through International Collaboration.* Palo Alto: Stanford University Press.

Index